SUPER NUKE!

A Memoir About Life as a Nuclear Submariner and the Contributions of a "Super Nuke" - the USS RAY (SSN653) Toward Winning the Cold War

CHARLES CRANSTON JETT

The opinions expressed in this manuscript are solely the opinions of the author and do not represent the opinions or thoughts of the publisher. The author has represented and warranted full ownership and/or legal right to publish all the materials in this book.

Super Nuke!
A Memoir About Life as a Nuclear Submariner and the Contributions of a "Super Nuke" - the USS RAY (SSN653) Toward Winning the Cold War
All Rights Reserved.
Copyright © 2016 Charles Jett
v3.0

Cover Photo © 2016 thinkstockphotos.com. All rights reserved - used with permission.

This book may not be reproduced, transmitted, or stored in whole or in part by any means, including graphic, electronic, or mechanical without the express written consent of the publisher except in the case of brief quotations embodied in critical articles and reviews.

Outskirts Press, Inc.
http://www.outskirtspress.com

Paperback ISBN: 978-1-4787-6642-1
Hardback ISBN: 978-1-4787-6649-0

Library of Congress Control Number: 2016904571

Outskirts Press and the "OP" logo are trademarks belonging to Outskirts Press, Inc.

PRINTED IN THE UNITED STATES OF AMERICA

It is with honor that I dedicate this book to my fellow plank owners and shipmates aboard the first Super Nuke, the USS Ray (SSN653) and to each and all of my fellow submariners who proudly wore their Dolphins and served with gallantry, dignity and honor in winning the Cold War.
Bravo Zulu!

TABLE OF CONTENTS

Foreword		I
Chapter 1	The Beginnings	1
Chapter 2	Welcome to Nuclear Power	11
Chapter 3	Nuclear Power School	24
Chapter 4	The Prototype	40
Chapter 5	Naval Submarine School	56
Chapter 6	Launching	74
Chapter 7	Fitting Out	93
Chapter 8	Fast Cruise	110
Chapter 9	Sea Trials	120
Chapter 10	Sea Trials - Bravo	129
Chapter 11	Sea Trials - Charlie	135
Chapter 12	Welcome to the Fleet	145
Chapter 13	The First Mission	154
Chapter 14	The Second Mission	179
Chapter 15	The Third Mission	197
Chapter 16	The Idea - Rejected	216
Chapter 17	Taking the Risk	224
Chapter 18	SSN Pre-Deployment Training	235
Chapter 19	Reflections	244
Appendix A	Albert L. Kelln	251

Appendix B Nuclear Physics, Reactor Principles,
            and Nuclear Reactor Technology 272
Appendix C Thermodynamics and Fluid Mechanics 280
Appendix D Radiation and Nuclear Waste 285
Appendix E The Super Nuke Plank Owners
            USS Ray (SSN653) 288
Acknowledgments 290

FOREWORD

Fifty years ago, the American public's attention was focused on the quagmire of the Viet Nam war and little attention was paid to something with a much higher priority. Quietly the United States was investing your tax dollars on something incredibly sophisticated and powerful: the US Naval Nuclear Submarine Force, whose focus was not on Viet Nam, but entirely on the strategic threat of the Soviet Union. The Polaris submarine force enjoyed public attention as a strategic deterrent, but little attention was paid—rightly so—to the "other guys"—the nuclear fast-attack submarines who covertly engaged the Soviet Union "up close and personal."

This book is dedicated to the men who proudly wore the gold and silver dolphins during this time in our history and served in the US Naval Nuclear Submarine Force—Polaris and fast attack. Your investment paid off! *Put modestly, these men and their magnificent machines quietly won the Cold War for you without ever firing a shot!*

Imagine what it would be like to be a crew member and officer who helped build and serve on the first and most modern nuclear fast-attack submarine ever designed to fight and win the

Cold War against the Soviet Union, and then imagine devising a way to successfully communicate to successive submariners the collective knowledge resulting from what we learned.

It was simply being in the right place at the right time with a fabulous crew.

Join me on that once in a lifetime experience that had a significant impact on winning the Cold War.

But first, here is an important word of caution and warning to those who have served or are serving aboard nuclear submarines.

You will find that some technical information and descriptions in this book might not be consistent with what you remember or what actually is the case.

This is intentional!

Such exact and detailed information is protected under the Atomic Energy Act of 1946 and is and will remain classified. Accordingly, while you may find something that you think is inaccurate and feel motivated to submit public "corrections" to what is described, understand that by doing so you may be divulging classified information for which you will be solely responsible. *This book has been cleared for publication by the United States Navy because it does NOT contain such classified information.*

There are thousands of men—officers and enlisted—who served with honor and distinction in our nuclear submarine force during the Cold War and proudly wore the gold and silver dolphins. All of these Cold War heroes who served in the United States Nuclear Submarine Navy will have had their own experiences and will be able to relate to some of the experiences described in this book.

All those who served aboard the nuclear fast-attack submarines—then and now—have unique and exciting adventures buried deep within the classified archives of their minds. Many of those missions were as exciting or more exciting than those experienced by the officers and enlisted men aboard the *Ray*.

However, only a very few will be able to relate to the experiences aboard that first operational 637 (Sturgeon) class submarine that was deployed against the Soviet Union— the *USS Ray* (SSN653). These men and that marvelous submarine were the pioneers, the "Daniel Boones," who blazed the trail with the most modern equipment and devised the sophisticated tactics and sonar developments that enabled subsequent nuclear fast-attack submarines to ultimately play a major role in bringing down the Soviet Union.

And only two of us can remember exactly how we collected, organized, and effectively communicated this information and growing knowledge to subsequent nuclear fast attack crews in preparation for their important covert Cold War missions. *To put this in perspective, any 637 or 688 class submarine that went through SSN Pre Deployment training before a covert mission in the 1970s or 1980s was getting trained from materials and experiences gained from the* Ray *in areas such as:*

- *The Geographic (Geo) Plot;*
- *The use of electronic surveillance equipment (AN/WLR-6);*
- *Range rating and more sophisticated application of sonar data.*
- *Specific tactical maneuvers.*

The purpose of this book is to tell the story about the source of this knowledge.

To begin, let's look at some fundamentals. We are already well into the 21st century and live nearly two and a half decades after the Cold War ended. If you are under twenty-five years of age, you never experienced the Cold War and might not even have a clue about what nuclear submarines are or what they did back in those "good old days." Accordingly, a brief and highly simplified history lesson—lasting only a couple of pages—is in order.

What Is a Nuclear Submarine?

Think of a nuclear submarine as a long steel tube, over a football field in length and ten yards in diameter, that can go very fast and undetected deep under the surface of the ocean anywhere in the world. It is something filled with exotic and state-of-the-art equipment and is the home and place to work for over one hundred men for months submerged without breaking the surface.

The word "nuclear" does not mean that the submarine has nuclear weapons. In fact, there were no nuclear weapons for the first nuclear submarines. Instead, the word "nuclear" means that the submarine has a nuclear or atomic powered engine. The

nuclear engine—the nuclear reactor—is the source of energy that drives the submarine around the world and powers the equipment to provide a high-quality living environment for its crew. When a nuclear submarine is built, its nuclear reactor has enough fuel to power the submarine completely for its entire thirty years of service. Think of buying a car that is full of a special type of gas—you can drive the car anywhere you want for as long as you want and never have to stop to fill 'er up!

That's what a nuclear submarine can do.

The Cold War

The Cold War is the name given to the relationship that existed between the United States (USA) and the former Soviet Union (USSR) from the end of World War II until the Berlin Wall was torn down in 1991. On a broader scale, the Cold War was a non-shooting conflict between two opposing economic systems: capitalism (USA) and communism (USSR).

While neither side engaged in a shooting war with the other, the period was marked by a high level of tension, distrust, and competition for supremacy. Each side had nuclear weapons of mass destruction and kept nuclear-tipped intercontinental ballistic missiles poised and ready to shoot at the other from land or from sea. Borders were closed and each side mounted enormous efforts to gather information about the other.

As a collective "weapons system," the US nuclear submarine force accomplished its mission with flying colors. Chairman of the Joint Chiefs of Staff and later Secretary of State Colin Powell said in 1992 on the celebration of the 3000th nuclear-powered ballistic missile submarine (SSBN) patrol: *". . . no one — no one — has done more to prevent conflict —- no one has made a greater sacrifice for the cause of Peace than you, America's proud missile submarine family. You stand tall among all our heroes of the Cold War."* What he didn't say that day but fully well knew was that while the *"SSBNs held a gun to their heads, the SSNs had 'em by the balls."*

This book simply points out that these submarines were highly effective in completing their designed missions. That is no secret, as General Powell said. The book tells in general terms "what" we did, but not exactly "where" or "how" we did it. Therefore, it is totally useless for any current or potential adversary to duplicate how those things were done. The book focuses on the activities of some of the incredibly competent men who participated as a team to make the Super Nuke a reality. Everything that is described in this book happened nearly fifty years ago.

Some may wonder why I have not included the 594 (Permit) class and earlier submarines as "Super Nukes." This is by no means a sign of disrespect and is in no way intended to diminish the role played by those fine submariners of earlier classes or their crews. In fact, those submarines carried on spectacular missions of which I'm fully aware. Their genuine adventures, however, have generally been confined to classified patrol reports. Any crew member is encouraged to write his own book describing the life aboard these fine submarines and, most important, the contributions they made and effectively communicated to the nuclear submarine force as a whole.

The reason is because my own first hand experience was with the 637 (Sturgeon) class submarines who were fitted out with the latest sonar systems and the best electronic intelligence-gathering systems ever put aboard a submarine — making them unequaled intelligence-gathering platforms and, accordingly, "Super Nukes."

And specifically, the USS Ray was the first—the "Daniel Boone," the pioneer, the original Super Nuke that created new sonar applications, devised new operating tactics, *and most importantly, communicated and shared these developments with fellow submariners in subsequent missions throughout the Cold War.*

This book is a memoir of an incredible six years as a junior officer in the United States Navy – but it is not an autobiography. Part of my naval service was quite normal, similar to that experienced by many of my classmates and others both before and after my time in the Navy. But the four years after my nuclear power and submarine training were remarkable: having the opportunity to serve with some of the brightest men in the Navy while building and serving on one of the most incredible machines ever designed by man. And the last two years in the naval service were unique: playing a major role in making a significant contribution to the US Naval Nuclear Submarine Force to bring down the Soviet Union.

It is a story that should be shared because I believe that any young man or woman—given the right opportunity—can and will do the same thing by preparing themselves for the future, finding themselves on the cutting edge of emerging technology, recognizing opportunity when it arises, and taking steps to seize that opportunity. I will try to accomplish this by taking you on an unclassified journey through those years as though you are with me.

The first commanding officer of the *USS Ray*, Albert L. Kelln, has joined me in sharing his experiences in the nuclear navy. His story may be found in the first appendix to this book.

I hope you enjoy this book and I hope that it gives you an understanding not only about the process of how someone goes through the nuclear power training program, but what a Super Nuke is, how it works, what it does, and the incredible quality of men who dedicated themselves to serving their country in our nuclear submarine force.

I was an active participant *but, more important, I was an eyewitness to their collective achievements.*

I hope this book serves to motivate young men and women to consider service in the United States Naval Nuclear Submarine Force.

Charles Cranston Jett
Chicago 2016

Chapter One
THE BEGINNINGS

I could see her in the distance, facing me and lying low in the water at the end of Pier 22 at the US Naval Base in Norfolk, Virginia. She appeared sinister—like a predator ready to stalk her prey. She was black, had no hull numbers, carried no flag, but I could see the two tall masts rising out of the top of her single coal black sail—a periscope and a radio mast. She was a submarine—a fast attack nuclear submarine—but not just any submarine. She was the *Super Nuke*—the first operational 637 class submarine specifically designed to face the Soviet Union. And she looked anxious to get underway and take care of business.

It was dark—two o'clock in the morning, and there was a slight fog which made the scene even more mysterious. The topside watch-stander returned my salute as I walked aboard and climbed down the hatch just after the black steel sail. After getting a cup of much- needed black coffee from the wardroom, I went back aft to assume my duty as the Engineering Officer of the Watch. After joining the three maneuvering watch-standers in the engineering control space (called "maneuvering") and settling on my stool, I said to the Engineering Watch Supervisor, "EWS, commence a normal reactor plant start-up."

With that command, we began the careful and well-rehearsed procedure to wake up the multi-megawatt nuclear power plant and prepare to go to sea. Shortly the Super Nuke would be operating under its own nuclear power. She was armed and ready for her first adventure—an encounter with the Soviet navy.

My road to Pier 22 and the Super Nuke began on the plains of the Dakotas, and the Navy wasn't an obvious choice for me. In fact, being from the western Dakotas, I had never seen the ocean. The only large body of water I had seen as a child was Lake Michigan and I was far more interested in Chicago's tall buildings than a fresh water lake. My only knowledge of the Navy was what I had seen in movies, and, on an annual basis, the traditional battle on the football field between Army and Navy. *And I was always an Army fan.*

My first six childhood years were spent on a large ranch in the far southwest part of North Dakota about twenty-five miles south of the town of Bowman and right on the North Dakota-South Dakota state line. My mother was a school teacher. She graduated from high school in LaGrange, Illinois at the age of sixteen, and then attended two years of college. When she was eighteen years old she was teaching in a one-room schoolhouse in the tiny town of Ladner, South Dakota in the far northwestern part of the South Dakota near the Little Missouri River.

She is the one who made me a strong supporter of early childhood educational programs such as "Head Start," because I benefited from something similar. My earliest memories include my sister, one year older than I, sitting with me in our kitchen on two small chairs facing a blackboard while our mother washed dishes. My mother would "teach school," as we called it, and that gave me the opportunity to learn the alphabet, count, add and subtract numbers, and read pretty well before I entered the first grade after we moved to South Dakota.

My grandfather, CB, helped with my early education as well while I rode with him in his Jeep as he herded sheep or generally inspected parts of the ranch. He taught me the alphabet—forwards AND backwards—as well as the names of the capitals of all the forty-eight states. By the time I entered the first grade I could also draw a map of the United States.

I wasn't too popular with my first grade teacher when she asked the class if anyone knew the alphabet. I raised my hand and rattled it off with a big smile—forwards and backwards. This didn't please her, and my reward was to be put in the back of the class and given a coloring book as she spent the next few weeks or months teaching the alphabet to the other kids. Looking back, it wasn't that I was any smarter than the other kids; it was just that I had the advantage of someone who was willing to spend some time teaching a preschool kid. It was an early introduction to education—I was the beneficiary of that, but the school system wasn't prepared to let me take advantage of it.

I grew up from first grade through high school graduation in Hot Springs, South Dakota, a small town in the far southwestern part of the state, with a population of about 5,000. During World War II, the western Dakotas were most likely the safest place on the planet, being almost equidistant from both the Atlantic and Pacific Oceans and far from any armed conflict. Certainly there was no danger of any foreign invasion from German or Japanese forces.

Reminders of war surrounded us, however, because the town was the site of a rather large Veteran's Administration (VA) Center, which was built in the early 1900s to take advantage of the area's warm mineral springs known for their therapeutic qualities. During my childhood, the VA Center was the largest employer in town and housed nearly 700 residential patients—mostly disabled veterans—with hospital and clinical care facilities offering outpatient services. Having the opportunity to visit with these World War I and Spanish-American War veterans was my first exposure to anyone connected with the US military. To this day I regret that I did not have the foresight, even at a young age, to take notes about the stories told to me by these American heroes. Their stories would have easily filled a book.

Throughout the time from my elementary schooling through high school, the Cold War began and tensions mounted between East and West—a state of tension that grew between the original allies of World War II and lasted from about 1947 through 1991. It was a conflict between two competing economic systems: capitalism in the West and communism in the East. As children, we were aware of what was termed the communist threat or the "Red scare." No one wanted to be called a "commie."

My mother often told the story about my rushing into the house in late June 1950, grabbing a sheet of paper and writing a letter to President Harry S. Truman. I wrote, "Dear President Truman: The North Koreans have invaded South Korea! Call the Army! Call the Navy! Call the Air Force! And quick! Love, Charlie Jett."

She helped me put the letter into an envelope, we addressed it to the president at the White House, and I rushed over to the post office (which was only a block away) and mailed it.

A few days later when we heard the news that the United States acted on a United Nations Security Council resolution to dispatch UN forces—mostly American—to the defense of South Korea, my mother told me that I came in the house with a rather long face, thinking that my letter had started a war. That was the first (and last) letter I ever wrote to the President of the United States. (I never received a response from President Truman.)

About sixty miles north of Hot Springs, near the town of Rapid City, is Ellsworth Air Force Base, a large strategic air command base that was home to squadrons of the mammoth B-36 "Peacemaker" bombers. I had taken several tours of the base and was always fascinated by the air force and flying. My father was a civilian pilot who owned a small Piper Cub airplane.

I enjoyed taking rides with my father in his small plane and we would occasionally fly up to the ranch south of Bowman, North Dakota where I would spend the summers driving a tractor and plowing the weeds in the fields that were fallow. There were also a couple of occasions when I asked my father to give my girlfriend and me a ride in the plane around the local area near Hot Springs. Little did he know that she and I were mapping places out near a large dam on the Cheyenne River where farmers had planted their secret watermelon patches. We used those maps occasionally to steal those watermelons.

When I was in middle school and in late May was preparing to go to the ranch for the summer, my father asked me if I wanted to ride up to North Dakota with Ed Anderson, a family friend. Ed had been a fighter pilot in the Pacific during World War II and owned and operated a crop-dusting service. I would be riding up to North Dakota in Ed's crop- dusting plane and that sounded like a lot of fun.

The flight up to North Dakota with Ed was unlike any experience I ever had. Ed, as a crop duster, was comfortable flying at an altitude of about 150 feet and following the contour of the land. We sped along in his crop duster, flying up the hills and down the hills and occasionally buzzing herds of sheep as the sheepherders waved their fists at us while their sheep scattered. It made me just a little more than woozy, and to combat a feeling of impending air sickness, I would occasionally put a dab of toothpaste in my mouth so the mint flavor would combat my increasing need to throw up. Fortunately, that treatment worked and we finally landed at the ranch after a two-hour ride – much like riding a "tilt-a-whirl" or an "octopus" in one of those traveling carnivals that occasionally would come to town.

After high school graduation, I was accepted as a student at Carleton College in Northfield, Minnesota with the intent of taking a pre-medicine major not only because it seemed like a noble profession, but even more because in my hometown the doctors were the ones who were making the most money. About a month after I had enrolled at Carleton, however, my parents told me that they were separating and that they wouldn't be able to afford four years of college for me. It was one of those things like *"Sorry about this, but have a nice life anyway."* That was quite a blow, because all during the time I was growing up, I was told that I should get a college education and that I could go to any school that I wanted. I was faced with a major problem and had to figure out what to do.

I recalled that during my first year in high school, the new Air Force Academy had opened in Colorado Springs, Colorado and my thoughts turned to the potential of seeking an appointment to the academy, because at least I could attend college as well as get involved with something of interest to me. I wrote a letter to the academy asking for information and they sent an attractive brochure to me describing the academy, the way of life of a cadet, and information regarding the process of seeking an appointment. On the train ride from Minnesota to visit my grandmother and uncle in western North Dakota for the long Thanksgiving weekend, I read the brochure several times and decided to begin the process.

During the weekend, I watched the annual Army-Navy game and rooted hopelessly for the Black Knights as a running back named Joe Bellino of Navy ran through them like a knife through butter on the way to a smashing Navy victory.

When I returned to Carleton, I wrote three letters: one to my local Congressman, E.Y. Berry, one to Senator Karl Mundt, and one to Senator Francis Case. I also contacted several people in my hometown of Hot Springs to provide references for me. One was an ex-governor of South Dakota and one was the nephew of Senator Case. A few weeks later, I received letters from all three. Both Senator Mundt and Congressman Berry granted me competitive appointments to the Air Force Academy. Senator Case, however, said that he did not have any openings for appointments to the Air Force Academy, BUT if I wanted a principal appointment to the Naval Academy, I could have it. A principal appointment meant that all I had to do was pass the entrance examination and I would be admitted. That sounded pretty attractive.

I thought about the opportunities for a few minutes and concluded that I would take the appointment to the Naval Academy – not because of a burning desire to become a naval officer, *but because it was farther away from South Dakota*. Little did I foresee the adventure that awaited me over the next ten years.

I had no problem passing the entrance examination and was accepted as a member of the class of 1964. On July 5, 1960, after taking the oath as a midshipman in the main court of Bancroft Hall at the United States Naval Academy, it was official! I was in the Navy and it didn't take me long to switch my football loyalty to Navy over my beloved Army team. Suddenly, that dreaded football running back, Joe Bellino, became a "good guy." Go Navy! Beat Army!

With a little effort, I adjusted to life as a midshipman, realizing quickly that if I learned the rules and regulations and followed them, I would have no problem. While I wasn't big enough to play Navy football, I was an excellent trumpet player, having played in a combo and dance band throughout high school. So I auditioned for the Naval Academy Drum and Bugle Corps and immediately found myself as the soloist for the half-time shows at Navy football games. That was fun – particularly playing solo during half time at the 1960 Army-Navy game where my favorite running back, Joe Bellino, who had just won the Heisman Trophy, starred again as Navy defeated Army!

A year later at the half time show at the annual Army-Navy game, I had the opportunity to serenade President John F. Kennedy—a loyal Navy fan. I stood in the center of the field with President Kennedy on the sidelines at the 50-yard line not far from where I was standing and played "Around the World" — blasting as loud as I could. When I finished playing, I brought down my horn and gave him a snappy salute. To my surprise, he gave me that big "Kennedy grin," saluted back at me, and then pointed his finger at me! When he did that, it seemed that the sky had suddenly become dark and there were two spotlights shining down – one on him and one on me. For about a tenth of a second there were no other people in the world – not the 110,000 people – not the national television audience – *just him and me*. It was surreal! I "connected" with JFK for an incredible moment – and that memory has remained and will remain with me forever!

Academically, I did well at the Naval Academy – the year at Carleton College had been excellent preparation. I enjoyed my time at Annapolis, making lifelong friends – and especially liked the structure and discipline of academy life. I didn't have time for a girlfriend; I focused on my studies.

My first exposure to the ocean occurred during the summer between my first and second year – "youngster cruise," as they called it. I was assigned to a destroyer based out of Norfolk,

Virginia and after I reported aboard we promptly went to sea, eventually spending a couple of weeks in gunnery exercises in the Caribbean.

My experience with a surface ship in the open ocean was not pleasant at all. I was introduced to "heavy seas" out in the Atlantic and for the first time I understood what "pitching and rolling" was all about. Seasickness grabbed me like a steel trap. There was no way to escape the motion anywhere on the ship – it always found me. I couldn't sleep below deck, and the only relief I could find – along with another seasick shipmate – was in the "flag bag" up on the bridge. The fresh air was helpful, but in general, the entire time at sea was miserable. It dawned on me that pitching and rolling on the sea is not something people generally experience on the wide plains of the Dakotas!

Eventually the summer cruise was over and I concluded that when it came time during my last year to make a service selection, I would pick anything but a ship that traveled on the surface of the ocean. That left few opportunities, and I vowed to study as hard as I could so I would have a high class standing and be able to select something that did not "pitch and roll."

I kept my vow to study hard and succeeded in getting my class standing high enough so that I would have a decent choice of service when the time came. Two and a half years passed by rather quickly and while a description of life and times at the Naval Academy might be interesting to some—even the subject of another book—that is not the point of this book. Therefore, I will jump to early 1964 when service selection was right around the corner – and my career as a naval officer was about to begin.

Chapter Two
WELCOME TO NUCLEAR POWER

My first class year (senior) at USNA was a good one, particularly in the world of sports, where Roger Staubach won the Heisman Trophy in football—we wound up ranked second in the nation even though we had lost to Texas in the Cotton Bowl. In other sports, Navy won the national championships in both lacrosse and soccer.

On the darker side, it was also an unusual and uncertain time in America, because just a few months before, President John F. Kennedy had been assassinated, war clouds loomed in Southeast Asia, and it seemed as though the Cold War with the former Soviet Union had entered a deep freeze. Scarcely a year before, we were engaged in the Cuban missile crisis —an event which brought this country close to all-out nuclear war.

All of the midshipmen were looking forward to the service selection process, which generally took place during the middle of the second semester. Finally, we would all find out what we would be doing after graduation: which part of the Navy, perhaps the Marine Corps, or in some rare circumstances, one of the other two branches of the military.

Service selections at the Naval Academy were based on class standing, with those at the top of the class having the first choices, and those below in order of their standing. My class standing was high, so I had a choice of virtually anything that I wanted to do, and was certain to avoid the fate of pitching and rolling on a surface ship. Becoming a naval aviator would not be one of my choices because of my eyesight, and service on a surface ship was out of the question because of my problem of getting seasick. The Marine Corps—a fine organization—required a lot of camping out in the mud as well as being fired on at close range, so that was a non-starter for me. The supply corps sounded boring, even though it had the potential of a promising future in retail or distribution; one of the different services (Army or Air Force) was not on my list because of my loyalty to the Navy. That left Nuclear Power—and specifically submarines—which had been my first choice all along. Submarine service provided almost two years of post-graduate education and service on a sea-going vessel less subject to causing seasickness.

My introduction to submarines, however, came as the result of a tragic accident. On April 10, 1963, the *USS Thresher* (SSN593) sank during a test dive in deep water about 220 miles off the coast of Cape Cod. All 129 ship personnel and shipyard workers perished. This was particularly tragic news at USNA because several of the officers lost in the accident were academy graduates; two of them were upper classmen during my first year. This was the deadliest submarine disaster in US history.

The *USS Thresher* was at the time the most advanced submarine in the world. It was the quietest submarine ever built and was the lead ship in a class of many such submarines scheduled for construction over the next few years. The principal dangers on any submerged vessels are fire and flooding, and to appreciate that danger aboard the *Thresher*, consider the following: Suppose you had a fire hose in your hand and someone turned the water on under the kind of pressure to which the hull of the *Thresher* was exposed near its test depth. The fire hose would shoot a stream of water over 750 feet in the air. That is a lot of force—and a lot of water.

The cause of the disaster is a topic still under debate, but it was most likely the result of a flooding casualty followed by a loss of propulsion power. The *Thresher* struggled in vain to surface, but quickly sank past its crush depth and fell to its final resting place 8,400 feet beneath the surface of the ocean.

It is a fact that such submarines have emergency escape trunks, and that submarine rescue vessels are able under some circumstances to rescue crew members from a sunken submarine. But such rescues are virtually impossible at depths below four hundred feet, where the water pressure becomes simply too great. Knowing that the average depth of the Atlantic Ocean is 10,900 feet and the average depth of the Pacific Ocean is about 14,000 feet, submarine rescues from the ocean floor are literally impossible. The pressure of the sea is unforgiving, and when the *Thresher* passed below its crush depth, the hull was crushed like an egg being smashed with a hammer. Death for the crew was instantaneous.

The loss of the *Thresher* weighed on my mind, but there are always risks in any branch of the military. I was confident that a redesign of the submarine would soon be accomplished and by the time I would be assigned to one of the boats (submarines are called "boats"), alterations would be made. This actually happened, and the *Thresher* Class boats—later called the Permit class—were modified to make them "sub safe." A later design, the Sturgeon (SSN637) Class, was designed from the keel up with all the sub safe features a part of the original design as well as the most up to date sonar and electronic equipment available. Eventually the boat to which I reported was a Sturgeon class submarine—the *USS Ray* (SSN653).

The Navy exerted pressure on midshipmen to enter nuclear power because of the high priority of building submarines to counter the Soviet threat. Many midshipmen, including myself, whose academic class standing was rather high, received a letter strongly urging them to enroll in courses—specifically, a course in nuclear power—that focused on the areas of science relevant to the nuclear navy. USNA scheduled many midshipmen to be screened for the nuclear navy through a series of interviews to be conducted by the Division of Naval Reactors (NR) in Washington, DC. These interviews for my class were scheduled for February 1964.

My turn to be interviewed came on a typical grey Maryland winter day. It was raining slightly and uncomfortably cold, but not freezing. The bus ride from Annapolis to Washington, DC along US 50 through the suburbs of eastern Washington, DC to the office buildings where NR was located was uneventful. There were twenty or so of us first-class midshipmen who were to be screened for the nuclear power program that day.

The buildings had been built early in World War II and were stark and uninviting. The offices were small, the walls undecorated, and the furniture was a typical Navy grey. The interview process, however, was efficient and brisk. Each of us was interviewed by two or three NR staff members, each neatly dressed in a civilian suit. None of the NR staff members had a smile to share with us as we proceeded through our individual interviews; instead, they were matter-of-fact, direct, polite, and above all, very, very smart. I was impressed—and close to being intimidated.

Each of the three NR staff members who interviewed me seemed to have his own agenda, and their questioning appeared to be very well-coordinated. One of them focused on my activities at the Naval Academy; one focused on my academic record and what elective courses I had taken; and the third asked specific questions about nuclear physics— especially the neutron lifecycle balance equation, "K eff," or "K effective." Fortunately, I knew all of the six factors in the equation and the fact that if "K eff = 1," this meant that the nuclear reactor was "critical" and was able to sustain a nuclear chain reaction.

During the day, I had also the chance to meet with two naval officers. I was told that they had been selected for command of nuclear ships (submarines and surface) and were attending a six-month period of temporary duty—Prospective Commanding Officer (PCO) training in the offices of the Division of Naval Reactors—before going to another PCO school in Groton, Connecticut for tactical training. They were very polite and quite willing to chat about their experiences. I remember them looking at me as though they knew that I was going to be experiencing an interesting but difficult challenge in the months ahead.

While the PCO program was affectionately called "charm school," it was a rigorous program of lectures, academic lessons, and examinations to ensure that each individual had thorough knowledge of nuclear power plant design and operation. I remember these encounters to be rather casual and not part of the formal screening process. Throughout the interview process, I remember being again struck by the fact that the premises were rather plain—not at all fancy or impressive as were some of the other government offices that I had seen in Washington. This place was very business-like and not intended to impress anyone.

When the three interviews were over and it came time for me to meet the admiral, I was led into a small office that had a large grey metal desk, with the admiral sitting at his desk facing the door as I entered. There was only one plain chair in front of the desk and the admiral told me to sit down (which I obediently did). He proceeded to ask me why a particular grade I had received in an applied mathematics course was not so high as some of my other grades. I told him that I had more difficulty with that course than some of the others, and then he abruptly told me to get that grade up and, at the end of the semester, to write a letter to him and let him know the final grade. I responded, "Aye aye, sir." (I should note that my applied mathematics grade actually dropped during the second semester because of my competing social life, and I never did write the follow-up letter to the admiral. Over the course of the next few years, I occasionally thought that I might receive a phone call or letter from him asking me about the grade, but neither that call nor a letter ever came.)

When the admiral was finished with me, he abruptly closed a folder which I assumed contained my files, and said, "That's all!" Then he waved his hand motioning me to leave and said, "On your way out, tell that WAVE lieutenant over there that she's the best-looking woman you've ever seen."

I was a bit startled, but rose quickly from my chair across the desk from the admiral, proceeded to the door, stopped, looked directly at the woman—who wasn't smiling—and said, "Ma'am, you're the best-looking woman I've ever seen." I don't remember whether or not the woman was or wasn't good-looking, and she didn't even look up from what she was doing at her desk to acknowledge what I had said.

With that I left the office, and my interview with the legendary father of the United States Nuclear Navy, Admiral Hyman G. Rickover, was over. I don't remember all of the questions he asked during our short conversation, but I do remember my promise to write a letter to him. I didn't have a clue about whether or not I had been accepted into the nuclear power program, but from what I had been told, if the admiral didn't just kick you out of his office, you had probably been accepted.

A couple of weeks later, another group of midshipmen who had gone through the selection interviews along with me went back to Washington DC for an interesting test with a pressure chamber. We were told that the pressure chamber test was intended to test whether or not our Eustachian tubes were suitable for service aboard submarines.

The Eustachian tubes are the tubes that connect the inner ears to the nasopharynx and enable equalization of pressure on both sides of the eardrum. When you fly in an airplane, for example, you occasionally have to "pop your ears" because of the lowering pressure in the airplane cabin—the pressure will equalize on each side of your eardrums if your Eustachian tubes are functioning properly.

The pressure chamber was located in a building that looked rather industrial in nature with a lot of pipes, valves, and instrumentation. The chamber itself was situated in the center of a large room, was sort of a cream color, and reminded me of an overgrown iron lung which I had seen many times in a polio hospital in my home town of Hot Springs. The chamber was not very large and accommodated about twelve midshipmen on two benches—six on one bench and six on the other, sitting shoulder to shoulder and facing one another. We were told that after the chamber was sealed, the pressure inside the chamber would be increased from atmospheric pressure (14.7 lbs/in^2) to 50 lbs/in^2 to a level over three times atmospheric pressure.

As the pressure rose, the inside of the chamber began heating up due to the compressed air and all of us had to "pop our ears" to equalize the pressure. A couple of the midshipmen were in considerable distress, either because they had poorly functioning Eustachian tubes or because they had a slight cold, which made their Eustachian tubes block up due to the swelling. All of us, however, successfully made it to the 50 lbs/in^2 level and we were left at that pressure for approximately five minutes. It was interesting that when we tried to speak at that pressure, our voices were much higher and rather "squeaky"—much like one sounds when breathing helium and attempting to talk. The most uncomfortable feeling for me was the much higher temperature of the chamber. The experience was like being sealed in a large coffin with a group of other people, with only a dim light turned on. When they reduced the pressure back to atmospheric, the chamber room became a bit foggy as the moisture in the air condensed and it became rather cool with the decreasing pressure.

Later on, I was told that the real reason we had to take the pressure chamber test was not entirely because of the need to test our Eustachian tubes. Rather, it was one test to discover if any of us was claustrophobic. While none of us indicated that the closed-in environment under high pressure was uncomfortable, I am certain that anyone with claustrophobia would have signaled that he wanted out of that chamber immediately!

I had the challenge of getting into the nuclear power program behind me and had completed my thesis (required of all midshipmen) early. I had seen first class (senior) midshipmen sweating writing their thesis into the late spring of their first class year and vowed that I would not put myself through that experience. So I wrote my thesis in the fall and had it submitted and graded before leaving for Christmas leave. I had always been interested in the strategic area of the Khyber Pass between Pakistan and Afghanistan and its history of being a pathway for invasions of such historical figures as Alexander the Great, Genghis Khan, and others. It was interesting then, but wasn't of much use to me until the Soviet Union invaded Afghanistan in December 1979. Then it became handy because I didn't have to refer to a map to find out where Afghanistan is located and I understood fully the rugged mountainous terrain between Afghanistan and Pakistan. The thesis was to show why the Khyber Pass has been strategic over the centuries, but the real reason I wrote the thesis was because of the mountains. Mountains don't make me seasick!

With little more than routine academics left, I decided to focus a bit on my social life, at least for my last semester at the Naval Academy.

First of all, it's time for some definitions from *Reef Points* — the correct name for what is referred to as the *Plebe Bible*. Plebes (first year students) at USNA must know what is in *Reef Points*— and this little book contains many bits of wisdom as well asan abundance of things that are quite irrelevant. In the world of dating at the Naval Academy, there were terms every midshipman had to know—remember, at that time USNA was still an all-male institution:

- **Brick** - *(n) the date who should have stayed at home; (v) to saddle a classmate with such a drag.*
- **Drag** - *(v) to escort; (n) young lady escorted.*
- **Queen** - *(n) opposite of a brick; from the ridiculous to the sublime.*

Up to this time at USNA, I had not dated at all except for the traditional Ring Dance at the end of the third year. The Ring Dance is a special event held in June Week of each year and is the time that the second class midshipmen (juniors) receive their class rings. It is customary to take your "queen" to the dance. I didn't have a "queen," but managed to find a "drag" who, fortunately, was not a "brick." That convinced me to shop around.

I had always been impressed with the very attractive young women who lived in the Annapolis area, and had wondered why so many of them were teachers from North Carolina. It wasn't that there were really a lot of North Carolina teachers; it just seemed that those I met from time to time were in that category. Later on in life I realized that a collection of over four thousand single, smart, and physically able young men might be an attraction for young single women looking to find a husband, but at that time I was clueless.

One Sunday afternoon, I was visiting the steerage in Bancroft Hall—a small area that had what was like a mini-soda fountain. I was getting some sort of over-rich sundae when a very attractive young woman came up to me and asked me for the time. I didn't have a watch, but gave her the approximate time. I didn't know why she was there—she didn't appear to be with any other midshipman—and I didn't bother to ask her if she was with anyone. Her name was Beverly, and we started chatting about music because she told me she played the piano and the organ. Since I had played the trumpet and had been a member of a dance band as well as the Naval Academy Drum and Bugle Corps, we talked about music. She told me that she was a junior at Western Maryland College. She seemed very nice and was quite engaging. We didn't chat long, but at the end of our brief discussion, she gave me her phone number and address and suggested that I give her a call.

That was simply too much to resist, so I immediately wrote a short letter to her after I returned to my room, realizing that I would probably never hear from her. To my surprise, I received a very nice letter from her later in the week suggesting that I give her a call and that we could perhaps get together. I called her that evening and set up a date for a movie in Baltimore. We seemed to hit it off right away, and after a few more dates over the next three weeks, I had completely fallen in love! She was my queen! I learned that she was a former "Miss Maryland"—obviously gorgeous. Western Maryland College was conveniently located not too far away, north and a bit west of Baltimore.

Over the course of the next few months up to and including my graduation in June 1964, I had focused almost entirely on her, to the detriment of my academic studies. That didn't seem so important at the time, because I had already been accepted into the nuclear power program, but it did cause me to drop my class standing by about fifteen places. Still, I graduated in the top three percent of the class, and when it became time to make my service selection, I chose nuclear power school, to be initially stationed at the school in Bainbridge, Maryland—close to my girlfriend's college. I also became engaged to her, but the promise I made to Admiral Rickover about sending him a letter still loomed in my mind. I didn't think he would have been pleased to hear about the advancement in my social life, nor would he have held me in high esteem knowing that my grade in the advanced mathematics course had not improved.

Immediately following graduation, I took a road trip along with my roommate in my new green Volkswagen beetle and we went down south to Alabama, out west to the Grand Canyon in Arizona, up to and through Wyoming's Yellowstone Park, and then to my home in the Black Hills of South Dakota. I enjoyed the visits with my family and a visit from my fiancée when she stopped in the Black Hills on her way to meet her mother, who was on vacation in Los Angeles. While she stayed only a couple of days, we had fun visiting Mount Rushmore and a few of the local tourist traps.

In late July I drove back to Maryland in my little green Volkswagen – thirty-one hours of nonstop driving with nothing but black coffee and sunflower seeds to keep me company— to report to nuclear power school in Bainbridge at the end of July, 1964.

My nuclear power career had begun!

Chapter Three
NUCLEAR POWER SCHOOL

Jack and I met up on the way to Bainbridge, Maryland at a Howard Johnson restaurant near Havre de Grace, a small town on the west bank of the Susquehanna River in eastern Maryland. Each of us had completed a marathon drive from the Dakotas: Jack from his home in Mott, North Dakota and me from Hot Springs, South Dakota. We were getting ready to report to the nuclear power school, but first wanted to have breakfast together and review what we knew about what was facing us. Jack was a USNA classmate and we had known each other since we entered the academy four years earlier and had even shared "youngster cruise" together, pitching and rolling on a destroyer in the Atlantic Ocean during the summer of 1961. Jack was one of the fortunate ones who was not susceptible to seasickness.

Over breakfast, we discussed the briefing materials and a short history about the Naval Training Center at Bainbridge. Some maps of the center had been provided, and we immediately located the Bachelor Officers' Quarters (BOQ) as well as the Officers' Club.

Nuclear power school is the first phase of training in preparation for duty as an officer aboard a nuclear-powered ship. The school is academic in nature, and its mission is to *"Train officers and enlisted students in the science and engineering fundamental to the design, operation and maintenance of naval nuclear propulsion plants."* In essence, nuclear power school teaches the relevant mathematics, physics, thermodynamics, electrical engineering, and chemistry that all come together in the design and safe operation of a nuclear power plant. For officers, the school is very rigorous, and for all practical purposes it crams approximately two years of difficult science into six months and teaches the equivalent of a master's degree in nuclear applied engineering.

The intent of the school was not to teach officers how to become nuclear engineers; rather, it was intended to teach them the necessary nuclear engineering and nuclear physics principles as well as relevant principles of other areas of science in order that they could understand fully how a nuclear power plant works and how to operate a nuclear-powered warship safely and with confidence. The next chapter will discuss how subsequent training at the prototype facilities enabled officers to actually experience the operation of a nuclear power plant and apply those principles. Looking back over the past fifty-plus- year history of the nuclear navy, this training has been remarkably successful: *there have been no nuclear accidents involving a US Navy ship in the history of the US Naval nuclear program.*

The first nuclear power training was established in 1955 at nuclear reactor testing facilities in Arco, Idaho and West Milton, New York where the initial training sessions were conducted by civilian engineers. The first formal Nuclear Power School was established in New London, Connecticut in January 1956 with a pilot course offered for six officers and fourteen enlisted men. Subsequently, in 1958, facilities were opened at the Mare Island naval shipyard in California and the Naval Training Center in Bainbridge, Maryland. The training center at Bainbridge was originally established in 1942 and over the years trained nearly a quarter million newly recruited Navy enlisted men in such areas as ordnance and gunnery, seamanship, fire fighting, physical training, and military drill. It was located on the northeast bluff of the Susquehanna River at Port Deposit, Maryland about forty miles east of Baltimore.

After breakfast, we drove across the river and to the training center, entered through the main gate, and went directly to the BOQ, which was quite easy to find. After parking my little green Volkswagen in the relatively small and crowded parking lot by the BOQ, I joined Jack in his 1959 Chevy convertible on a tour around the training center to get our bearings. The buildings had been built in haste—as most were at the time of World War II —on the site of an old abandoned school. By the time the nuclear power school was established in 1958, the facilities were quite decrepit, but still functional. The buildings housed the classrooms as well as small rooms that served as living and study spaces for the officers enrolled in the program.

The Officers' Club was a place where officers could go and relax with cocktails, dinner, breakfast, and entertain guests from time to time. The old and quaint town of Port Deposit was just down the hill from the bluff on the shores of the Susquehanna

River, but as we later learned, there was little to do in the town except to go occasionally to a bar that served cold beer and good but overcooked chili.

After reporting at the main office of the training center, we went back to the BOQ to move into the rooms to which we had been assigned. The rooms were stark and each room had a desk and a refrigerator—an excellent place to store our favorite bottled beer, Pabst Blue Ribbon, which we bought by the case at a "beer store" across the river on US Route 40 near Havre de Grace. There was a small closet and a single bed. While stark, the rooms were functional for studying and sleeping, which was all we were there to do.

Life at nuclear power school turned out to be a rigorous five-day work week with the days crammed full of academic classes and the evenings spent studying, leaving little time for social activities. Making frequent visits to my fiancée at Western Maryland College even during weekends was impossible because of the heavy academic load, and this started to put a strain on our relationship.

My fiancée complained about the lack of frequent visits and didn't seem to understand the level of intensity to which those of us in the nuclear power program were subjected. This was distracting and interfered with my studying, making me wonder whether I had made a mistake. This feeling was compounded through my discussions with one of the officers, Don Tarquin, who lived in the room next to mine. Don and I developed a strong friendship and I would be sharing an apartment with him during the next phase of the program.

Don was a USNA graduate out of the class of 1958 and was senior to Jack and me, but quite willing to spend time with us. Over the course of the program, Don carefully explained what life was like aboard a submarine. He had served on a diesel submarine, the *Sterlet*, before being selected into the nuclear power program, and had also served in a non-nuclear capacity as weapons officer aboard the nuclear submarine *Sargo*.

Don did not encourage me to get married to someone who appeared to be highly dependent on my consistent presence. After all, Don explained, submarines go to sea, often for long periods of time, and a naval wife must be able to endure frequent separations. Don could see that there would be problems down the road, and he wasn't afraid to share that fact with me. That made me worry more.

The Course of Study

In order to help you get a grasp of the scope of the academic course load, it might be useful to provide some examples of what was being taught and how it applies directly to a nuclear propulsion plant. I will summarize each course—and if you are interested in learning more about nuclear physics, reactor technology, and reactor principles, you can check out Appendix B, where I've tried to make it simple to understand. If you want to learn more about thermodynamics and fluid mechanics as they apply to a nuclear power plant, check out Appendix C. Appendix D has information about radiation and nuclear waste.

Nuclear Physics, Reactor Principles, and Nuclear Reactor Technology

If you have ever wondered how a nuclear submarine can go to sea for months at a time just by splitting atoms as a source of energy, then nuclear physics, reactor principles, and nuclear reactor technology are the places to start. Don't be intimidated, because while the technical details might be rather complicated, the overall concept is rather simple.

The whole purpose of a nuclear propulsion plant on a submarine is to extract energy safely from nuclear fuel and to provide power to propel the submarine, as well as to provide electrical power for the boat. Nuclear physics, reactor principles, and nuclear reactor technology are the branches of science that show how to accomplish part of this process. The challenge is to safely split the nucleus of an atom, capture the energy released by converting mass into energy, and put that energy into some form so that it can be converted to electricity and mechanical power to propel the boat.

All that happens in a nuclear reactor is a controlled chain reaction, where atoms of the nuclear fuel U^{235} are split by slow-moving neutrons and some of the mass of the U^{235} atom is converted into energy in the form of kinetic energy and radiation. The kinetic energy is captured in the form of heat, which is generated by friction when the fission fragments of a split U^{235} atom slow down, and this heat is transferred to the primary coolant, which then travels to the steam generator to create steam which powers the main turbines and turbo generators.

Thermodynamics and Fluid Mechanics

Thermodynamics is the branch of physics concerned with heat and temperature and their relation to energy and work. When the nuclear reactor has finished adding energy to the coolant, thermodynamics and fluid mechanics take over to convert that energy into something useful—such as electricity, or propelling the boat through the ocean with the main turbine.

Fluid mechanics is the study of fluids (liquids, gases, plasmas) and the forces that act on them. This part of physics is important to a nuclear power plant because the energy created by the nuclear reactor moves around both the primary and secondary systems in the form of a fluid (water or steam). Further, the entire submarine and its related cooling and control systems are highly dependent on fluid mechanics: the hydraulic systems that control the boat's movements, the systems that control the boat's internal atmosphere, and the design of the submarine's shape itself, because the entire submarine exists and operates in a fluid.

Thermodynamics and fluid mechanics as taught at the nuclear power school are limited to the use and application of the energy created in the nuclear reactor. The application of thermodynamics, and especially fluid mechanics, was taught at the naval submarine school, where the principles were applied to the operating submarine as opposed to the nuclear power plant.

While the process of converting high-energy steam into mechanical and electrical power is a well-known subject, one of the unique features of a pressurized water reactor used by nuclear submarines is that the power generated by the nuclear reactor is controlled NOT by moving control rods in the nuclear core, but by what is called "steam demand." If the submarine needs to increase speed, which requires more power, all that needs to be done is to open the throttle to the main turbine for more steam. The system, through the temperature of the water, automatically increases reactor power to a level just enough to meet the demands for the required power. For details about how this happens, you can read Appendix C.

Mathematics

The mathematics course at nuclear power school didn't pose much of a problem for me, since I had considerable math exposure through advanced courses at USNA. The focus of the course was to teach the math concepts necessary to understand nuclear physics and to quickly solve problems that involved logarithms, extrapolation, the ability to determine the levels of radiation caused by potential coolant leakage, and the like.

One particular concept was a bit confounding for me, and I'll admit that I never really became adept with the equations, nor did I ever have the need to apply them to anything practical. This was the concept of "buckling."

In essence, "buckling" deals with neutron leakage. Geometric buckling is a measure of neutron leakage, while material buckling is a measure of neutron production caused by nuclear fission minus absorption of neutrons by a wide variety of elements within the nuclear core. In a steady state condition where the reactor is critical, geometric buckling is equal to material buckling. I thought that was a nice concept and interesting in its own right, but again, I had no occasion to apply it. If it is confusing to you, then welcome to the club!

One of my counterparts did encounter the concept in later life, but not during his career in nuclear submarines. After he had left the Navy, he was a graduate student in nuclear physics at the University of Minnesota. During his oral examination after writing his thesis for his PhD degree, the first question asked of him by the review panel was, "What is buckling?" He smiled and recalled the days at nuclear power school when he and I had pondered what buckling was all about and then gave his answer—successfully! I'm not sure if there was any other occasion after he received his PhD degree that he had the opportunity to use the concept, but it is widely used in nuclear reactor core design. I am certain, however, that if he later had an opportunity to ask a question of a graduate student in a similar situation, he would have asked, "What is buckling?"

Chemistry and Materials

Chemistry and materials are important in a nuclear power plant because of one important problem, and that problem is *corrosion*. The primary loop of a nuclear power plant must be constructed of some sort of material that can withstand high temperatures and high pressures. Additionally, whatever this material might be, it must be resistant to corrosion at those conditions of temperature and pressure.

The reason for needing corrosion resistance is because as the corrosion products flow through the primary coolant loop, they eventually pass through the reactor core many times and are subjected to bombardment by a high density (called a high "flux") of neutrons. When neutrons bombard the atoms of different elements, sometimes the result is the creation of a new element that can be highly radioactive and dangerous. Such an element, for example, is Cobalt^{60}. This is the element which is often used in medicine in radiation treatments for cancer. It is the element used effectively in such advanced medical instruments such as the "Gamma knife."

Co^{60} is a synthetic radioactive isotope of cobalt with a half-life of 5.3 years, meaning that its radioactivity will decrease by half in 5.3 years. It is produced artificially in nuclear reactors. Co^{60} is largely the result of multiple stages of neutron activation of iron, so when atoms of iron coming from corrosion of metal in the primary coolant loop pass through the nuclear reactor core, they can be and often are transformed eventually into Co^{60}.

The addition of Co^{60} to the primary coolant loop is not a welcome event because corrosion products often tend to congregate and settle in parts of the coolant loop that are not subjected to high coolant flow. Such parts include the vents of main coolant pumps and are often the sources of high gamma radiation in reactor compartments when the compartments are entered for maintenance purposes. Corrosion also takes place in the secondary steam loop, but the dangers of radiation there are not present because there is no primary coolant in the secondary loop.

To counteract the effects of corrosion, corrosion-resistant materials are used for primary coolant loop piping—materials such as Inconel—from a family of austenite nickel-chromium super alloys. Inconel alloys are oxidation- and corrosion-resistant materials well-suited for service in extreme environments subjected to pressure and heat. In all cases, however, careful monitoring and maintenance of primary loop chemistry— particularly pH – are priorities. In the secondary loop, careful monitoring and maintenance of the chemistry of the water in the steam generators are maintained.

Part of this maintenance is the use of highly distilled and virtually pure water that, when needed, is injected into the primary coolant loop as well as the steam generators. Chemistry samples are routinely taken from primary coolant and monitoring systems such as instruments that detect high salinity are placed throughout the secondary system, and alarms are sounded if these detectors sense salinity levels out of the ordinary.

Electrical Engineering

The electrical engineering curriculum focused primarily on how motors and generators work, the principles of magnetic amplifiers, and the principles of electric and electronic circuits. This was relevant to the submarine because of the need for the nuclear power plant to generate electricity, utilize various electric motors for pumps, and the design and use of electronic circuits to service in nuclear instruments such as detectors and meters. At USNA we had covered electrical engineering quite thoroughly, so this part of the curriculum was not particularly difficult for me.

Radiation and Nuclear Waste

Since we would be operating a nuclear propulsion plant, it was imperative that we understood the various kinds of radiation, the concept of minimum dose levels, and the principles of shielding against the hazards of radiation. The perils of radiation, primarily the dangers of radiation from spent fuel and nuclear waste, are covered in more detail in Appendix D.

The kinds of radiation created by nuclear reactors include Gamma, Alpha, Beta, and neutron radiation—all the kinds of radiation which we might encounter not only aboard a submarine, but at the upcoming prototype sites as well. We learned about the dangers of each kind of radiation, where each kind originated, and the steps that we should take to monitor the level of radiation to which we might be exposed. I learned for the first time that the level of radiation to which any individual is exposed on a very sunny day is greater than the radiation that he/she might experience aboard a nuclear submarine, even when close to the reactor compartment.

Exercises included determining the kind of radiation that individuals—even the general population—might encounter after a nuclear plant accident. What kind of radiation would they encounter? What would be the effect of wind and rain? What would be the dosage level to be expected at different distances from the reactor accident?

The entire course on radiation was pointed at reactor safety—not only what to do in case of a nuclear accident, but how to work on equipment and valves that were exposed to primary coolant. During this course, I began to gain an appreciation for the level of detail and the very strict processes that were followed by the Division of Naval Reactors (NR) that Admiral Rickover ruled with an iron fist. With nuclear waste, there is no margin for error, and Admiral Rickover insisted on specific procedures as well as exceptionally high standards of quality not only in the operation of naval nuclear power plants, but in the materials that were used in their construction. As a result of these strict procedures and quality control measures, there has never been a nuclear accident involving a US Naval nuclear power installation. *Never! And that is true up to this day!* My admiration for Admiral Rickover and what he had created was justifiably sky-high.

Power Plant Characteristics

This course was fun, mainly because it was a chance to apply what we had learned. The course included presenting various scenarios that we might encounter in nuclear plant operations and then having a discussion about what sort of indications and/or alarms we might expect.

Some of the scenarios included:

- *Starting up a nuclear reactor plant*
- *Main coolant pump operations*
- *Reactor scram procedures*

Each of the scenarios required us to think through what was going on and to apply what we had learned about how the reactor plant and engine room operate as a unit. It was an excellent preview to the upcoming second phase of our nuclear power training: six months at a nuclear reactor plant prototype. It was also an excellent preview of many of the processes we would be following on the submarines to which we would be assigned, as well as the drills we would be having on a regular basis—some actually overseen by Admiral Rickover himself!

Over the course of the six-month program at Bainbridge, we made frequent trips to Baltimore because we all signed up for and became members of the new Playboy Club downtown. This was fun, especially because of the pricing scheme: everything was a buck and a half—drinks, food, etc. This fit our meager budgets quite well. Rather than having a membership card, the unique feature of the Playboy Clubs back then was that you were given a key—an actual key—that had the Playboy Club logo on it. You could put it on your key chain to give you some sort of "status," whatever that meant.

We also took advantage of Maryland crabs in some of the fine and not-so-fine Baltimore seafood restaurants. We learned how to eat crabs when they would pile them up on your table on top of some paper covering, and give you a small wooden hammer and a knife. There was a definite process to eating Maryland crabs, and each of us learned that process well. Of course, Maryland crabs were incomplete without the addition of at least one pitcher of ice-cold National Bohemian Beer!

On Sunday evenings, Don and I had a ritual. We would drive to Port Deposit just down the hill from the training center on the banks of the river and have some of that overcooked chili at the bar whose name I can't remember, and which I think is no longer in business.

With the academic part of nuclear power completed, it was time for us to proceed with an introduction to the practical application of the academics. I was ready for the change, but my fiancée wasn't. She didn't understand why I had to go so far away and told me that she expected me to visit her at least every other week — a situation that I knew was physically and mentally impossible. My doubts and worries had now turned to thoughts of an exit strategy.

My now good friend, Don Tarquin, had become my mentor. Later on in his career, Don became the commanding officer of a fast-attack nuclear submarine, the *USS Drum* (SSN 677) as well as becoming the prospective commanding officer (PCO) training officer for the Pacific submarine fleet. He has always been an honorable and excellent officer in the tradition of the naval service—someone who has always had my highest respect. To this day he refers to me as "Ensign Jett"—a title that I accept with warmth and pride.

Don and I decided to share an apartment together at the prototype, and soon after we finished nuclear power school, we selected as our prototype site the Knolls Atomic Power Laboratory site in West Milton, New York. After packing our meager belongings, we traveled together in tandem up to Saratoga, New York—just ten miles east of the prototype site—and finally found an apartment in the little town of Gansevorrt, just five miles north of Saratoga. My friend Jack Grant and another classmate, John Nurenberger, took the apartment next door to Don and me. We were a "foursome," just right for playing some bridge, which we did regularly during the months ahead.

Phase two of the nuclear power training program—The Prototype—had begun.

Chapter Four
THE PROTOTYPE

The winter of 1964-1965 was cold! Both Jack and I came from the Dakotas and when we mentioned that to anyone, virtually everyone would comment about how severe the winters are out there in the West. But from my experience, the South Dakota winter was mild compared to that in the Saratoga area. Not only was there deep snow, but it was cold, cold, cold!

My little green Volkswagen didn't seem to mind, however, and no matter if it was below zero when Don and I would get ready to leave for the prototype at five o'clock in the morning, the little engine started right away. Don's fabulous GTO convertible was not so accommodating, however. Sometimes that big engine would turn over slowly, and maybe would start…and maybe it wouldn't. We learned to depend on the German automobile engineering!

The dictionary defines a "prototype" as a "first, typical or preliminary model of something, especially a machine, from which other forms are developed or copied." This very basic definition describes what the nuclear power prototypes are all about. The prototype experience give officers in-depth practical experience at a real operating nuclear power plant to apply what they learned during phase one of the training program. The requirement is to "qualify" on each watch station normally manned by an enlisted member as well, as to qualify as an engineering officer of the watch (EOOW) in charge of overseeing the operation of the power plant during actual operations.

The prototype experience consisted of two phases. Phase one was approximately three months of classroom instruction, where each system of the nuclear power plant was described in detail. Here we learned the mechanical systems, the reactor system, control systems, electrical systems, and the like. Classroom instructors included the Navy officer who was in charge of the program, as well as engineers who actually designed each of the systems.

When I finished nuclear power school in Bainbridge, Maryland, I had a choice of two prototypes as part of the second phase of the nuclear power training. One site was in Arco, Idaho—one of the original sites—and the nuclear power plant at that site was the prototype for the plant that is aboard the *USS Enterprise* (CV 65)—an aircraft carrier. The other site was in West Milton, New York and that site contained two prototypes, the D1G prototype for the plant that is aboard the *USS Bainbridge* (CGN-25)—a nuclear-powered guided missile cruiser—and the S3G prototype for the plant that is aboard the *USS Triton* (SSRN 586). There was a third prototype in Windsor Locks, Connecticut—unavailable to us at the time because it was under an overhaul process—and that was the S1W prototype for the plant that was aboard the *USS Tullibee* (SSN 597).

Of the two available sites and the available power plants, I chose the S3G prototype in West Milton, New York because it was specifically a submarine, albeit a rather unusual submarine. In reality, it didn't make much difference which prototype one chose; the programs of instruction and qualification process were the same. I also chose that site because it was closer to where my fiancée was going to college, although it would not take long for that relationship to sour.

Don Tarquin and I finally found a suitable apartment near the little hamlet of Gansevoort, New York just a few miles north of Saratoga Springs. Two of our friends, Jack and John, shared the apartment next door. After moving in and becoming comfortable with our new living quarters, we checked out Saratoga Springs and found several interesting places such as the various spas, the horse racing track, nifty pizza joints, and most interesting, a women's college—Skidmore College. We decided to keep the latter in mind, even though I was on the trailing edge of being engaged. (I was slowly coming to reality.)

On our first visit to the prototype site about twenty miles from our apartment—accessible only on back country roads—I had my first experience with nuclear alarm systems. As I walked through the main gate, an alarm went off and I was immediately approached by a guard.

"Excuse me, sir!" said the guard. "Are you wearing a radium dial watch?"

"Yes," I replied.

"Well," said the guard, "you can't go on site with that watch. You'll have to remove it, leave it here with us, and pick it up when you leave."

Don smiled. Apparently he knew that something like this would happen. He later told me that radiation alarms on board submarines are sensitive enough to detect the radon gas given off by even the smallest radium dial watches.

I took off the watch and said, "Sorry. I didn't know this would be a problem."

The guard said, "Happens all the time, sir. No problem at all. You just have to leave it here. Regulations, you know."

I left the watch with the guard and when we departed later that evening, retrieved it back for safekeeping at home. Curiously, since that time I have never worn a wristwatch.

Don and I checked in and attended the first day's orientation, which would describe our first three months' routine. The routine was as follows:

- *Be on site from 6:00 a.m. until at least 6:00 p.m.;*
- *Use the time from 6:00 a.m. until 8:00 a.m. for study in our individual cubicles;*
- *Classroom instruction would take place from 8:00 a.m. until 4:00 p.m.;*
- *Use the time from 4:00 p.m. until at least 6:00 p.m. for study in our individual cubicles.*

Welcome to the Nuclear Navy, I thought. Twelve-hour days… that sounded pretty heavy, but at least we were told that we would have the weekends off. However, we could always have access to the site for study purposes at any time 24/7 during the weekend. That sounded like fun, indeed! And totally inconsistent with being engaged.

We met our nuclear power instructor for the orientation promptly at 9:00 a.m. Our instructor was LCDR Jack McNish, an experienced nuclear submariner. He told us a bit about his background, which included service aboard two nuclear attack submarines (SSNs), one being the *USS Halibut* (SSGN 587). The *Halibut* was a one-of-a-kind submarine, originally designed to carry the long-range surface-to-surface Regulus missile.

McNish was not a large man. He was about 5'10", was stocky, and his red hair was crew cut. To me he looked "all Navy," neatly dressed in his khaki uniform and sporting a bright orange belt instead of the traditional khaki belt that we wore. Later on he explained the "belt system" at the prototype.

- *Orange belts were worn by the lead prototype classroom trainer;*
- *Red belts were worn by prototype instructors who supervised training in the prototype itself;*
- *White belts were worn by those trainees who had qualified as engineering officers of the watch on the prototype (our goal was to become a "white belt");*

- *Khaki belts were worn by the trainees who had not yet qualified on the prototype (that included all of us sitting in the classroom).*

McNish was a very smart officer and an outstanding instructor. It was clear during the course of his lectures that he knew what he was talking about in great depth. Over time, I much preferred his classroom instruction over those of the various civilian design engineers from General Electric—the company that managed the Knolls Atomic Power Laboratory (KAPL). He had the ability to make complicated things very clear so that when studying the materials later, the subject and systems he described were easy to learn. I consider him to be one of the best instructors I ever had, from high school through the Naval Academy. He did not discuss anything about nuclear submarine operations. He was "all nuclear business" when teaching about the S3G prototype.

As you might suspect, a nuclear power plant has many "systems" that consist of steam piping systems, hydraulic piping systems, electrical power systems, electrical control systems, and the like. Not only did we need to know the purpose of each of these systems, we needed to know how to draw the diagrams for each and how each component of the system worked. Finally, when we were in the second phase of the training—the actual operation of the nuclear power plant—we had to trace the systems so we knew where the piping was and where each component was located. This was excellent training and preparation for me, since my first duty station was construction of a fast-attack nuclear-powered submarine.

Part of the routine in the classroom phase was to take a periodic walk through of the S3G nuclear power plant. The S3G nuclear power plant was located up on a small hill away from the classroom buildings, and access was through a building between the classrooms and the plant itself. In the building next to the prototype, there were many large pumps, all making sort of a whining sound as they pumped water through the cooling system of the plant.

The power plant itself was housed in a containment facility that was a relatively long tube shaped somewhat like a hot dog, but painted a light shade of blue. We had to climb a steep set of stairs up to the air locks of the plant for access. The air locks consisted of a double pressure door system—much like those found on a submarine—where you had to go through the outer door, close the door, then equalize the pressure, and then open the inner door to go into the plant itself. The air pressure inside the plant was slightly lower than outside, because if there were to be a leak inside the containment facility, any vapor or gas would be contained within the containment facility and not leak to the outside.

Just inside the airlock was a special compartment, constructed solely for the purpose of the prototype. This was a "water brake" —a large piece of equipment that could be used to simulate the real shaft of a submarine that was used to turn the screws (propeller) of the submarine to give it motion. Obviously the prototype was not at sea, so it needed something to provide some sort of resistance for the shaft when operating the main propulsion turbine. The water brake provided this resistance, and a way to dissipate the energy created by the power plant.

When the main turbine was engaged to simulate getting underway on a boat, the shaft turned the water brake and this generated heat. The heat was removed by a cooling system, which circulated water in and out of the water brake and cooled the water in an adjacent cooling tower. I thought at the time that there was a lot of energy being wasted by just heating up a water brake and then transferring the heat to the atmosphere. I was thinking of the possibility of having the shaft turn some sort of generator where the power could be converted to electrical and perhaps stored in some fashion. But that was a thought that was most likely unrealistic and not cost-effective.

Immediately upon entering the power plant after the water brake room, three large control panels were prominent on the left of the large room. These were the propulsion control panel, the electric panel control bench board (EPCB), and the reactor plant control bench board (RPCB). (These panels had different names on the S5W Westinghouse plant, like the one on the submarine to which I was ultimately assigned.) An operator was stationed in front of each panel, which contained multiple levers, switches, and many indicator dials. The Engineering Officer of the Watch (EOOW) stood behind them alongside one of the trainees who was in the process of qualifying on that watch station. Surprisingly, for an operating plant, the noise level was quite low—one could even speak in a normal voice and be easily heard.

You could see the myriad cables for electrical wiring all tucked neatly in bundles along the bulkheads, and insulation (most likely asbestos, I thought) covering the steam pipes that led to the turbo generator and to the main turbine.

Access to the lower-level engine room was through a hatch in the deck and down a vertical steel ladder. The lower-level engine room contained several detectors that monitored the salinity level of the condensed water from the steam, and several large pumps. Of particular interest to me were the main feed water pumps that pumped feed water that had been condensed from the steam back to the steam generators in the reactor compartment. These feed water pumps were steam-operated instead of electrically powered. That was unusual, we were told, because the main feed water pumps on all other submarine reactor plants were electrically driven. These steam-driven feed water pumps would play a major role in one of my watches as an EOOW after I had been qualified.

The reactor compartment, located just forward of the upper level engine room, did not seem to be so cluttered as both the upper and lower engine rooms. Within the reactor compartment were the steam generators, which turn water into steam to drive the ship. In this design, the steam part of the steam generator had a special sight glass built into it so a person standing watch could monitor the water level inside the steam generator. This sight glass, too, would play a future role in one of my EOOW watches.

Near the starboard bulkhead in the reactor compartment was a place known as the "alley." This was a row of two large cabinets, each about six feet in height and about ten feet in length, facing each other with about three feet separating them from one another. Each of these cabinets was full of electronic equipment that consisted of reactor control components such as circuits for the reactor control rod drive mechanisms, etc. It was a complicated place simply because of the overwhelming visual image of so many electronic components in one place, but when one understood what was the purpose of the controls, things became a whole lot easier. Nevertheless, I always became a bit nervous when going into "the alley."

My strategy and approach to the prototype training was to visit the S3G power plant daily and spend time looking at each component at all levels. Given the challenge to eventually be able to comprehend and understand every piece of equipment in the plant, my process was to tour the plant daily and become very familiar with the equipment and surroundings through the repetitive visits. Eventually, I reasoned (and hoped), I would be able to diagram the entire plant from memory; it turned out that I was right!

Both Don and I had settled into a good routine of out of bed at 4 a.m., at the prototype for study at 6 a.m., classes from 8 a.m.-4 p.m., study and touring the plant 4 p.m.-6 p.m., and back to the apartment. It was relatively easy to maintain this routine even though we were in the middle of experiencing a nasty upstate New York winter. To pass the time in the evenings, we would watch television and some of the old shows that I'm sure you might remember, such as *Mission: Impossible, The Man From U.N.C.L.E., Bonanza, Gunsmoke*, and the like.

Together with two other naval officers who were my USNA classmates also attending the prototype, Don and I began to play bridge in the evenings. A typical evening would include a couple of quarts of Genesee beer and a couple of pizzas from D'Andrea's pizza in downtown Saratoga.

I had broken my engagement with "Miss Maryland" a couple of weeks earlier. Don and our other two bridge partners decided that it was time that the four of us check out Skidmore College. After all, they reasoned, Skidmore was a women's college and we were all single guys. They pondered for a while about what strategy they could use to introduce all of us to some interesting and attractive girls, and after taking a vote about how to proceed, designated ME as the one to find four girls for us.

I remembered from *Reef Points*, the USNA "Plebe bible," a short piece that we were forced to memorize when we were plebes. It is titled the *"Qualifications of a Naval Officer"* —supposedly written by John Paul Jones—and the first lines are as follows:

> *"It is by no means enough that an officer of the Navy be a capable mariner;*
>
> *He must be that, of course, but also a great deal more.*
>
> *He should be a man of liberal education, refined manners, punctilious courtesy, and the nicest sense of personal honor."*

I didn't recall that the "Qualifications of a Naval Officer" also required one to be adept at convincing four college women to go out on a date with four naval officers attending nuclear power school, but I decided to give it a try—nothing ventured, nothing gained—and of course, I would exercise "punctilious courtesy."

After the TV show *Bonanza* on one Sunday evening in February, 1965, I picked up the phone and called the Skidmore College general number. When the operator with a very business-like voice answered, I said, "Can you connect me with the dormitory?"

She responded quickly, "Which dormitory do you want?"

Without much thought I muttered the word, "Skidmore . . . " and the operator asked, "Skidmore Hall?"

I immediately said, "Yes!"

Then she asked, "What floor?"

I just pulled a number of a hat and said, "Third!"

She said, "Thank you!" and after a few seconds I heard the phone ringing.

After about three rings, I heard a female voice answer "Hello?"

I then launched into my unrehearsed pitch.

"Here's the deal," I said. "There are four of us—all naval officers—who are looking for four SENIORS who might be interested in playing bridge with us."

I continued, "We are willing to come to Skidmore Hall so you can check us out and decide whether or not you want to join us. If you do, then we will take you to D'Andrea's Pizza, feed you, then to our place where the eight of us will play bridge for a couple of hours. What do you think?"

To my surprise, the girl answered, "Sounds good! I'll round up three of my friends."

"We want SENIORS only," I said.

She replied, "No problem—I'm a senior and I'm thinking only of my friends who are seniors. Can I call you back when I find out if any are interested?"

"Certainly," I said. And with that we said our goodbyes and I waited for her to call back.

About an hour later, the phone rang and she was on the line.

"I found three more," she said. "When do you want to come over?"

I was elated! I told her that we could come by Skidmore Hall around 6:30 p.m. the next day, a Monday. She agreed, and I told Don and my two classmates. They collectively applauded my success! Personally, I was a bit surprised.

Promptly at 6:30 p.m. that Monday, the four of us parked in front of Skidmore Hall and went inside looking for our potential bridge players. To our surprise, my contact and her three friends were waiting for us in the lobby. Each was quite attractive, and it didn't take us long to decide that we should head out together for some pizza. When we were at D'Andrea's Pizza, we decided to take out the pizza instead of wasting time in the restaurant. We took the girls back to our apartment in Gansevorrt where we spent about three fun hours playing bridge. The girls were delightful and all of us agreed to give the event an encore.

Over the next few months, the eight of us got together at least on a weekly basis, enjoyed each other's company, and became good friends. Ultimately, two of the girls, Lois and Susie, whom we met that evening, married two of my friends. My roommate, Don, married Lois; John married Susie; both Jack and I escaped with our bachelorhood intact.

At the end of three months, the classroom phase of the prototype training program was over. It became time to "qualify" on the actual nuclear power plant. To "qualify" on a nuclear power plant meant, for an officer, that he must "qualify" on each watch station normally manned by enlisted watch-standers. There were watch-standers for a wide variety of places and pieces of equipment, such as lower level engine room, reactor plant control bench board, electric plant control bench board, and the like.

Since the nuclear power plant was running 24 hours per day and 7 days per week, our class of officers was divided into various watch duty sections that lasted eight hours each. We would be required to be on site and stand watch for those eight hours, all during our process of qualifying on the reactor plant.

Qualification was accomplished by standing watch at a particular station for a period of sessions—generally three—or until the officer could be signed off by a qualified watch-stander as knowing how to man that particular watch station. Officers were required to qualify on each watch station normally manned by enlisted personnel, and then to qualify as an Engineering Officer of the Watch (EOOW) after standing a number of watches under the supervision of a qualified EOOW. That was supposed to ensure—and it generally did—that the newly qualified EOOW really knew what he and those members of his watch team were doing.

After all of the qualification signoffs had been accomplished, final qualification was achieved by facing a review board consisting of the various training officers from the prototype site and the commanding officer of the site itself. The review board could ask questions about virtually anything relevant to the nuclear power plant, including what sort of maintenance was underway at the present time.

During my own qualification, for example, there was maintenance being done on one of the steam-driven feed pumps. A problem had occurred that was difficult to diagnose. My review board was scheduled to begin at 8 a.m., so very early—around 5 a.m.—I went into the nuclear power plant and spent time with the technicians who were diagnosing the problem of the feed pump and attempting to fix it.

When my review panel began, after I had answered a couple of rather simple questions by members of the panel, the commanding officer asked, "Can you give me an update on the feed pump problem in the plant?"

I was ready for that question, since the problem had finally been fixed about thirty minutes before my review board began and I was fairly certain that no one on the board was aware of the resolution of the problem. I explained the problem of the feed pump control system to the board, drew a diagram of the control system on the blackboard, and showed that there was a slight oil leakage coming from a bellows. I then reported that the bellows had been replaced, the feed pump control system tested, and it was back online.

That was news to the entire review board, and the commanding officer basically qualified me on the spot. I had done what had been taught to me by LCDR Jack McNish—and that was to always be able to describe a problem by drawing a diagram, and always be prepared to answer questions about equipment that was having problems—particularly equipment that could jeopardize the boat's ability to get underway or operate in any fashion. That lesson would serve me well when I was assigned to new construction.

Following my qualification as an EOOW, I spent approximately two more weeks at the prototype site before this phase two of the nuclear power training was over.

Now it was time to take a couple of weeks' leave and then go to Groton, Connecticut to the US Naval Submarine School and spend the next six months learning all about submarines.

Phase three of the nuclear power training program had begun.

Chapter Five
NAVAL SUBMARINE SCHOOL

Say the wise: How may I know their purpose?
Then acts without wherefore or why.
Stays the fool but one moment to question,
And the chance of his life passes by.

What this one "Law of the Navy" means is that sometimes an opportunity comes your way, and if you pause, you might lose the chance of a lifetime.

So it was, I thought, with most of my submarine school classmates. All of the careful thought, the meticulous planning, the thoughts about being home with the family, the worry about extra pay all of that came into play when first duty station selection day came and went. Careful thought and planning were the norm for most if not all of the members of my submarine class. For me? Well . . . what wasn't obvious to others turned out to be my *unintended stroke of genius*. To put it mildly, I "got lucky and won the prize!"

Before embarking on any details of the submarine school, let me give you a quick primer or refresher about how a submarine works. Let's make it simple.

Assume you have a tin can—aluminum would work just as well. It's open at one end and on the other end there are two holes, but they have plugs in them. The holes/plugs are the "vents."

You put the tin can into water with the open end at the bottom. First, however, you put a little water into the can, but only until it is about half full. That makes the can float with the closed end up.

You remove the two plugs in the vents at the top of the can. Air rushes out of the can and it starts to fill up with water. When the weight of the can gets to be more than the buoyancy of the remaining air in the can, or when the can is full of water, the can will submerge. When it does, you close the vents. The ideal situation is when the can nearly full of water is slightly heavier than the buoyancy provided by the remaining air in the can.

Your can is submerged! It's a little submarine!

When you want the can to rise to the surface, with the vents closed, you put some air into the can to blow out some of the water from the bottom. The can then becomes lighter than the water and your submarine surfaces!

That's all there is to it!

Now let's return to reality and check out the Naval Submarine School.

That Groton, Connecticut may be described as a "Navy town" is an understatement, for almost fifty per cent of the town's population is employed either by the US Naval Submarine Base or by a major contractor for building submarines, the Electric Boat division of General Dynamics.

You often hear about the "submarine base in New London," but contrary to popular belief, neither the Naval Submarine Base nor the Naval Submarine School is actually in New London, Connecticut, even though the main gate at the base says, "Naval Submarine Base New London." They are located in Groton, Connecticut, across the river on the east side from New London. And the river is the Thames River, with the "Th" pronounced the way you say the word "through," not the way you pronounce the famed river in London, England. The only submarine facility to be located in New London was Submarine Squadron 10 and the *USS Fulton*, a repair ship based at State Pier in New London.

The submarine base did not look like a "school" at all—it looked like a real Naval base in the real Navy, or at least what I thought a "real Naval base" should look like. It was certainly unlike the Naval Training Center in Bainbridge and the prototype in West Milton, New York. There were real Marines at the main gate! After entering the main gate, you could see the "lower base" on the left down by the river, and the unmistakable silhouette of both diesel and nuclear submarines moored at the various piers.

Since I had never been close up to either a diesel or nuclear submarine, I made a left turn and drove down the hill to the lower base. The diesel submarines were located at the piers on the eastern part of the lower base, downstream from the nukes.

While they looked like submarines, it was obvious to me that they were really surface ships that had the ability to submerge, and later I found that this was a correct description.

The nuclear submarines, on the other hand, looked like *real submarines*. In fact, they had an ominous appearance, most of the submarine (nearly 90%) being underwater and only a small rounded part of the hull and the unmistakable sail with the masts and periscopes rising high above the sail, and the "fairwater planes" that stuck out horizontally from the small sail like wings.

About midway from the southernmost part of the lower base to the northernmost part stood a strange-looking tower that appeared to be over one hundred feet tall. It looked like a silver-colored round cylinder with a spiral staircase attached to the outside and a slightly larger cylinder surrounding the top of the tower. Later I learned that this was the famed "escape tower," where sub school trainees were given an opportunity to do a buoyant ascent from a pressure chamber at the bottom of the tower through a column of water approximately one hundred feet tall to the top of the tower. Knowing that I would be participating in this didn't bother me at all since I was an excellent swimmer, had no problem with my ears during pressurization, and had considerable experience in scuba diving. I knew, however, that it would be a challenge for others.

At the far upstream part of the lower base was a building that had a sign that said "Submarine Development Group Two." I found out later that the SUBDEVGRUTWO, as it was designated, was the place that had been originally established to develop tactics for fast-attack nuclear submarines to seek out and destroy enemy submarines. That sounded interesting to me, but it wasn't until my tour of duty after serving aboard the *Ray* that I had any direct contact or meaningful business with the development group.

Leaving the lower base, I immediately saw the large building that housed the Bachelor Officers' Quarters (BOQ). After parking my little green Volkswagen, I went into the BOQ, registered, and was assigned to my room which would be my home for the next six to eight months.

Compared to the Spartan accommodations of the Naval Training Center at Bainbridge, the sub base BOQ rooms were spacious. Each of us had a separate room complete with a single bed, a sitting area, a large desk for studying, a closet, and a shared shower/bathroom with an adjacent room. My adjacent roommate was my fellow "Dakotan," Jack, who had been with me in nuclear power school as well as the prototype. Jack and I would hang out together throughout sub school, including pursuing social activities in and around the Groton/New London area.

Compared to nuclear power school and the prototype training, sub school was a breeze. There were 104 officers in my class—a typical class size for those days when nuclear submarine construction was at its peak. The curriculum was focused not on the nuclear aspects of submarines, but on basic knowledge of how submarines work, how they operate, and the kind of knowledge needed to support the many systems typically found on both diesel and nuclear submarines. In a sense, the principles were much like the "tin can" example, only a lot more complicated.

Specific course areas included the following: principles of submarines, principles of sonar, basic submarine tactics, submarine weapons systems, submarine weapons (mainly torpedoes), and communications.

There were practical training exercises conducted as well from time to time. One type was training in the various submarine attack center simulators where teams could simulate not only periscope approaches on surface ships, but sonar approaches as well. These tactical trainers were set up the same way as a real submarine control center, with a periscope and the control panels that controlled the weapons. They were, however, for the most part simulators using diesel submarine equipment, and the tactics that were taught were those that had been used during World War II. At the time I thought that perhaps these techniques were a little bit outdated, but I didn't come to a full realization of that fact until we encountered "up close and personal" the Soviet navy during initial operations on the Super Nuke.

Another type of practical trainer was the diving trainer. These devices simulated the control room of a diesel submarine. They were used to teach the principles of how to submerge and surface a boat, but just as important was the practice they gave to trainees about how to "trim" the boat.

In a submarine, there are various water tanks, forward and aft, where water may be transferred back and forth through pumping systems that would shift the weight either forward or aft. The trick was to balance the boat—that is, to make sure that the boat was nearly level and the weight forward balanced the weight aft, and therefore, balanced the boat.

The angle of the boat—whether the bow was up or down—was called the "bubble" (based on a bubble indicator visible to the diving officer). When the "bubble" was at zero, the boat had a "zero bubble." The goal of the diving officer when trimming the boat was to adjust the trim so that the boat had a "one degree down bubble"—that is, the bow of the ship was slightly lower than the stern. This provided an ideal trim situation to ensure that the boat could stay submerged. When this condition was met, which was often difficult to do, the diving officer would announce, "Trim satisfactory, sir! One degree down bubble!"

These training sessions in the attack centers and diving trainers were held quite often, and many of the trainees had trouble, especially trimming the boat. When the time came to go to sea for practical training, the trainees would have a chance to trim a real submarine at sea. I found that to be a lot of fun to do.

On the social scene, I had basically put my experiences and memories of "Miss Maryland" in the past. New London was the home of Connecticut College for Women (Conn College) which was located across the Thames River up on the hill beyond the United States Coast Guard Academy. Jack and I wasted no time in exploring opportunities and looking for ways to introduce ourselves to prospective dates. Our competition at Conn College appeared to be the "Yalies"—the college kids from Yale University about sixty miles to the west of New London. They weren't much competition though, because while the Ivy League is impressive, they didn't stand much of a chance with the blue and gold Navy guys and the romantic nature of the submarine service.

Soon we had made a few contacts and began having some fun in around the area, including a few trips to Boston for the great seafood.

We had one opportunity for an "at sea" experience around halfway through the six-month submarine school program. Both Jack and I were assigned to the *USS Sea Robin* (SS 407), a World War II diesel submarine. Our two weeks at sea included practicing surfacing and diving the submarine, where each of the sub school officers had an opportunity to perform as diving officer—giving the command to dive from the bridge, sounding the alarm, coming down from the conning tower and closing the hatch, and checking the "christmas tree" to see if all the lights were green. The "christmas tree" is a panel that shows the position of certain valves that are critical to the integrity of the boat and MUST be shut in order to prevent flooding when the boat submerges. Green lights indicate that the valves are closed, so the diving officer looks first at the "christmas tree" to see of all the indications were green. If they are green, he shouts, "Green board, sir!" If they are not all green, the next step is to give the order to "Emergency Blow" to bring the boat immediately back to the surface.

As the boat begins to descend, we would "Blow negative to the mark!" This is a classic diesel submarine command that has a very specific and useful meaning. There is a tank near the center of the submarine which, when the submarine is on the surface, is full of water and provides negative ballast to help the boat submerge when the vents for the main ballast tanks are opened. As the boat submerges and is under control, the command,

"Blow negative to the mark!" is given to remove most of the water from the negative tank, thereby assisting in bringing the now-submerged submarine to neutral buoyancy. The negative tank is not completely emptied, but "blown to the mark" to ensure that air bubbles do not escape when submerging—the air bubbles might help an enemy find the submarine. After the negative tank is "blown to the mark," the diving officer announces to the conning officer, "Negative tank blown to the mark, sea valve shut, request permission to vent negative." When the conning officer grants permission to vent negative and the tank is vented, the process of establishing an appropriate trim— pumping water fore and aft to reach a one degree down bubble—can be achieved for submerged operations. When the diving officer announces, "Trim satisfactory, sir!" they would generally surface the boat and the next student would take his turn.

The first week at sea between New London and Hamilton, Bermuda where we pulled in for the weekend was my second experience in rough weather. There was a blizzard in Connecticut/Massachusetts—probably a "Nor'easter"—and the experience had two effects on me: 1) it made me recall my time in the flag bag on a destroyer during my midshipman cruise, and 2) it made me renew my vow NOT to select a surface ship where I would be subjected to the pitching and rolling caused by the ocean waves. Sadly, it also made me aware of the fact that diesel submarines are, in fact, surface ships that have the ability to submerge now and then. I dreaded the thought of being assigned to one of those.

After a fun weekend in Hamilton, Bermuda, we set sail south for

Puerto Rico and occasionally conducted at sea drills. One fun event along the way occurred when we stopped out in the middle of the Bermuda Triangle in very calm—almost glassy—seas, posted a guard with a high-powered rifle on the sail of the submarine to protect us from any sharks, and went swimming. The water was crystal clear, but even when we would dive into the water from the top of the sail, there was no danger of hitting the bottom. We were right over the Puerto Rico trench, which is on the boundary between the Caribbean Sea and the Atlantic Ocean and the water depth at the nearby Milwaukee Deep is over 28,000 feet deep—over five miles (the deepest part of the Atlantic Ocean). We never saw any sharks!

After our arrival in Puerto Rico, we spent a couple of days being tourists before our flight back to submarine school When we arrived back in Groton, Connecticut after the Nor'easter had blown out to sea and after all the snow had melted, it was back to class. We knew that selections of our submarines to which we would be assigned were coming up in a few weeks, so many of us began to think about exactly what kind of submarine we would like to choose. I had bad dreams about being assigned to a diesel boat.

Shortly after the diesel cruise, I had the opportunity to ride a new Polaris submarine on her "bravo trials." These are the trials where the submarine is taken to its test depth for the first time and put through various "angles and dangles," which meant diving at a very steep angle, ascending at the same steep upward angle, and testing various systems at test depth, including a test emergency blow where the submarine surfaces from test depth rapidly and virtually "jumps out of the water" when reaching the surface. I will describe this in greater detail later in the book when I cover the *Ray's* sea trials.

In late April, I was invited to be the host for the sponsor of the *USS George C. Marshall's* commissioning ceremony. I asked my New London girlfriend, Jackie King, to accompany me as we met Mrs. George C. Marshall at the New London/Groton airport and took her to the boat to rest and relax in the boat's wardroom before the ceremonies began. The wardroom is the room reserved for naval commissioned officers to provide a dining room as well as a place to relax. Officers eat their meals together in the wardroom and can come and go as they please either to relax or do their work without interruption.

We sat, had coffee, and chatted with the delightful Mrs. Marshall, talking mainly about her famous husband, who was the Army Chief of Staff during World War II, the father of the "Marshall Plan" created when he was Secretary of State under President Harry S. Truman. She had a sparkle in her eye and was actually quite moved knowing that this submarine was being named after her husband. General Marshall, for all his accomplishments, was a very humble man, one who valued his privacy and rarely sought public acclaim. It was impressive to hear her speak about how he felt about working with Presidents Roosevelt and Truman; he liked President Roosevelt and simply adored President Truman because he was so decisive.

Mrs. Marshall talked about what it was like to be with the general during World War II, and described that when he would come home from his work, he could put an entire world war out of his mind for a few hours and focus on her and her interests. She didn't mince words about how the general detested General Douglas MacArthur. It was quite interesting hearing a diminutive 80+ year-old woman calling that five-star general an "arrogant SOB."

We enjoyed chatting with Mrs. Marshall for nearly an hour as she enthusiastically expressed interest in the world of the submarine as well as sharing her past experiences with us. We felt a little sad having to end the conversation with her, as she was called to join the commissioning group for the ceremony. It was a genuine pleasure to meet her. That would be the only time we would spend with her, as the ceremony included her participation later in a reception and a dinner. Unfortunately for us, someone else would be escorting this delightful woman back to the airport.

After Mrs. Marshall departed to join the commissioning ceremony as the sponsor of the boat, I reflected on our conversation and one of my favorite memories came back to me. While I never met General Marshall (I truly wish that I had met him—he's one of my heroes), I did have a chance to meet President Truman, albeit very briefly.

I had always been lucky when it came to chance meetings with interesting and sometimes famous people. When I was in Chicago several years earlier to take my physical examination for the Naval Academy, I stayed at the Sherman House. The morning after my physical, I was in the lobby and decided to walk up the staircase to the restaurant, and on my way up the stairs this nice gentleman with a cane and a big smile was coming down the stairs. We arrived at the first landing at the same time, and he said, "Well, good morning, young man!"

I stopped for a moment, thinking that the gentleman looked a lot like President Truman. He smiled broadly, extended his hand, and said, "I'm Harry Truman! What's your name?" I told him my name and he asked, "What are you doing in Chicago?"

I told him that I had just taken my physical examination for the Naval Academy and he said, "That's great, son! Good luck to you there! It's a fine place!"

With that he told me he was on his way to a meeting, and continued down the stairs. I don't remember that he had any security people or Secret Service agents with him at the time, but times were different then. As I continued up the stairs to the restaurant with feelings of astonishment and awe, my mind flashed and I realized that I hadn't been so quick on my feet as I should have been. I wished that I had asked him if he had received the letter that I sent to him back in 1950 and told him that the North Koreans had invaded South Korea and that he should "Call the Army, call the Navy, call the Air Force—and quick!" I knew that he wouldn't remember, even if he had seen that letter from a little elementary school kid from South Dakota, but it would have been fun to apologize to him for my starting the Korean War. I smiled and chuckled to myself and went on to breakfast.

Back to the present and at sub school, my interest, through my conversations with Don Tarquin at the prototype, was to select a fast-attack submarine (SSN). Through another friend I contacted the commanding officer of a fast-attack submarine, the *USS*

Flasher (SSN 613), which was under construction at the Electric Boat shipyard. His name was Commander Kenneth M. Carr and he agreed to give me a tour of the submarine in the shipyard to see not only how it was being constructed, but the benefits of being a part of a pre-commissioning crew of a SSN as well. Commander Carr was certainly someone who knew a lot about nuclear submarines; he was one of the officers on the first crew of the first nuclear submarine, the *USS Nautilis* (SSN 571).

The *Flasher* was an interesting boat in that its original design was exactly the same as the *USS Thresher* (SSN 593) which had tragically sunk during a test dive in the spring of 1963. Construction of Thresher-class submarines was halted after that disaster, and a complete redesign of the class was completed. This design included the *Flasher*, so its construction was delayed significantly in order that all of the revisions and redesigns could be incorporated into the boat. When commissioned, the *Flasher* became the first of the "Sub Safe" boats resulting from lessons learned from the *Thresher*.

Commander Carr was an outstanding tour guide and an exceptional recruiter because after my two-hour walk through of the *Flasher,* I was ready to select SSN new construction without giving any other kind of submarine—nuclear or otherwise—even a second thought. During our tour, he told me privately, "Charlie, if you want to really understand how a submarine works, then build one! There absolutely is no substitute for new construction. It's hard, but it is a rare opportunity—if it comes your way, then seize it!" This was the potential opportunity that, at least for me, was the one referenced in the "Law of the Navy" quoted at the beginning of this chapter.

That did it for me! With our submarine selections coming up in just a few weeks, I began to dream about such an opportunity. If I had been ranked number one in the class, I would have selected SSN new construction.

I hit a major bump in the academic road, however, shortly before the boat selection time arrived. I was doing quite well—probably not so well as Admiral Rickover would like, but adequate and certainly within reach of making an early boat selection. My class standing at the time I took the tour with Commander Carr was somewhere in the middle of the top third of the class of 104.

But then I hit the bump in the road. I developed a rather nasty case of the flu and missed a little over a week of academic classroom work, just around the time some major tests were coming up. I was well enough to attend classes one day before taking the tests, but when I took them, my results were terrible. My class standing suffered because of it, dropping me from being in the top third of the class—a comfortable place to be to make a good boat selection—to a level where I was the "top man" . . . but, unfortunately, *in the bottom half of the class.* I was number 53 out of a class of 104.

The interesting thing was that boat selections were based on class standing for the top half of the class, but for the bottom half of the class, you had to draw a number that would give you your place in the selection list. So, my outlook wasn't so good and I worried about it because the only way my selection priority could go was down. I didn't realize just how far down I would drop until the night came when the bottom half of the class had to draw numbers for their place in the list of choosing boats. As the top man in the bottom half of the class, I got to make the first pick. And the number I picked? *Of course . . . 104 . . . the very last pick.* I was crushed and began to come to grips with the fact that after sub school, I would be spending the next few years being seasick on some foul-smelling diesel relic.

A few days before "selection day," the list of boats where billets were open was published. As expected, there were 104 billets available, and the choices included virtually every king of submarine imaginable. There were operating fast-attack submarines (SSNs), fleet ballistic missile submarines (SSBNs), diesel submarines, SSBNs in overhaul, SSNs in overhaul, and two billets for one new construction fast- attack submarine, the *USS Ray* (SSN 653) which was being built at the Newport News Shipbuilding and Drydock Company in Newport News, Virginia. It was scheduled to be commissioned over a year away from the time of the selection process. I thought my prospects were grim indeed!

One of the benefits of submarine service is that officers who are assigned to commissioned submarines are given what is known as "sub pay," a significant monthly amount that is comparable to "hazard pay" or "flight pay" that naval aviators receive. An important thing for those making a choice was to consider the fact that in order to receive "sub pay," the key words were "commissioned submarine." If the boat to which you were assigned was not yet commissioned (as is the case with all boats in new construction), then you would not get sub pay unless you were already qualified in submarines. That meant whoever picked the *USS Ray* under new construction, unless he was already qualified in submarines, would not receive sub pay until the *Ray* was commissioned — at least a year away.

There was a rumor spreading, too, that new construction was very hard work—dreadful for married people—and was something to be avoided at all costs.

I actively encouraged the spread of this rumor.

When the evening came for the class to meet in the sub school auditorium and make boat selections, a couple of friends joined me at the Officers' Club to chug down a few beers. That being done, when I arrived at the sub school auditorium to learn my fate, I was feeling rather fuzzy. In fact, I was probably drunk.

The first pick was an operating SSN out of Groton, the *USS Tinosa* (SSN 606). The next few picks were operating SSBNs and, surprisingly, a couple of diesels. Each SSBN has two crews—a "gold" crew and a "blue" crew. These crews alternate serving on the boat every three months. When a SSBN comes in from a sixty-day patrol, the crew on board gets relieved by the other crew and has some time off for relaxation and training. Three months later they take over the boat again, prepare to go to sea, and take another sixty- day patrol. That kind of duty seemed to appeal to the married officers. Blue crews and gold crews of operating SSBNs were going like hotcakes. Fast-attack boats were a different story. The few that were available included some operating SSNs, which were selected within the top fifty choices, and just a few commissioned SSNs that were in shipyards for overhauling or refueling.

When the selection numbers got down to the high nineties, there were a couple of SSBNs in overhaul left, a couple of diesel boats, one SSN in overhaul, and TWO billets for the *USS Ray* (SSN 653)! My hopes were rising and I didn't dare breathe. When number 102 was chosen—a diesel boat—all that was left were two billets for the *USS Ray* (SSN 653). I sobered up quickly! Most of the officers who were left in the auditorium were chuckling as Owen Brown, a classmate, made his selection. When it became my turn, I stood up and proudly selected the *Ray,* and my selection was greeted with a loud applause and many cheers, mostly in jest.

Little did anyone know, however, that fate had awarded me the biggest prize of the evening: an unforgettable adventure! I got exactly what I wanted, smiled silently to myself, and slept very well that night! For in just a couple of weeks I would be reporting for duty at the Newport News Shipbuilding and Drydock Company in Newport News, Virginia.

The adventure was about to begin—I was about to meet "Super Nuke"!

Chapter Six

LAUNCHING

My first memory after arriving in Newport News, Virginia late in the evening of April 18, 1966 was that Julie Andrews did not win the Academy Award for best actress for her performance in *The Sound of Music*. She had won the award the previous year for *Mary Poppins* and everyone seemed to think that she was a shoo-in for another Oscar. But that didn't happen. Julie Christie won the award for her performance in a movie called *Darling* . . . both the movie and the actress have long been forgotten by me.

Newport News is among the larger cities in Virginia and is located on the southeastern end of the Virginia peninsula on the northern shore of the James River. It is part of the Hampton Roads metropolitan area and many of its residents were employed at the time by the Newport News Shipbuilding and Drydock Company as well as the joint US Air Force and US Army installations of Langley Air Force Base and Fort Eustis Army Base. Accordingly, the city's economy was very much connected to the military and with the recent contracts for submarines and aircraft carriers, the economy was booming.

I had arrived in the city late in the afternoon and had the opportunity to drive around and look at different parts of the town for an apartment, which I would definitely need for the expected year-long period of building the submarine. Given that I would be living on the meager salary of a lieutenant (junior grade), the amount of rent being asked by many of the more attractive places was a bit more than I could afford. I would probably have to look for a roommate.

For the short term, I settled into an older hotel in the downtown part of the city within walking distance of the entrance gate to the shipyard. The hotel wasn't shabby, but it wasn't the Ritz either. The rooms were rather small, but quiet and clean and would be my home for at least a few days. After grabbing a pizza from a nearby Italian take-out restaurant, I settled in for the night to watch the Academy Awards show and to think about what was in store for me the next day.

As a member of the US Naval Nuclear Power program, my experiences thus far were virtually identical to my classmates' and to those who went through the program before and after me. But upon arrival at the shipyard in Newport News where they were building the newly designed 637-class submarines, everything changed. As more time went by up to and including my next duty assignment back at the Naval Submarine School, the experience became one of a kind.

Bright and early the next morning, I proceeded down to the far eastern end of the Newport News Shipbuilding and Drydock Company piers to where an old two-story Navy barge was moored. The barge reminded me of the "stewards' barges" at

USNA— painted grey and generally unattractive. The *Ray* was not a commissioned US Navy ship, so going aboard did not require any Navy protocol. Technically I was reporting to the Pre Commissioning Unit (PCU) for Hull 572, which was under construction. Boats under construction at Newport News are given hull numbers, and the *Ray* was "Hull 572." I went aboard the barge and into the main deck area, where there was a large open space with some cubicles scattered here and there as makeshift offices. To greet me was the Executive Officer (XO) of the PCU, LCDR Bob Harris.

After a brief chat with the XO and filling out the necessary papers to formally become a member of the unit, I met with the PCU Commanding Officer, Commander Albert L. Kelln. (Since ship commanding officers are referred to as "Captain," I'll refer to Commander Kelln as "Captain Kelln" for the rest of this memoir.) Captain Kelln greeted me warmly and told me that I would be taking the responsibility for the Electrical and Reactor Control (E & RC) Division in the engineering department. Owen Brown, my fellow sub school graduate, would be assuming the role of communications officer. That suited me just fine, because I wanted to get involved in the construction of the nuclear power plant.

Captain Kelln was an exceptional nuclear-qualified officer. He had experience serving aboard three nuclear submarines, the *USS Skate* (SSN 578), the *USS Shark* (SSN 591), and the *USS John C. Calhoun* (SSBN 630) before coming to the *Ray*. To illustrate Captain Kelln's level of competence and his knowledge of the way shipyards work in building and overhauling naval ships, he was personally selected by Admiral Hyman G. Rickover to serve as chief engineer for the refueling of the giant nuclear-powered aircraft carrier, the *USS Enterprise* (CVN 65). This was an enormous task and one that Captain Kelln accomplished with efficiency, earning him a great deal of respect from Newport News Shipyard management. Fresh from refueling the *Enterprise*, Captain Kelln was selected for and assigned as commanding officer of the *USS Ray* (SSN 653).

I then met the chief engineer of the PCU, Lt. Reid H. Smith. Reid was a graduate of USNA in the class of 1961 and turned out to be an extremely competent engineer who was easy to work for and a pleasure to call a shipmate. He was very well-organized and literally a workaholic who loved his job. Under his leadership and that of Captain Kelln, the *Ray* (Hull 572) set a record for the least amount of time it took from laying the keel of the boat to commissioning. After having a long chat with Reid about the role and responsibilities of the E & RC Division Officer during the new construction period, we donned out helmets and our steel-toed shoes and walked through the shipyard up to the shipways where Hull 572—the *Ray*—was comfortably perched.

She (naval ships/boats are referred to as "she") was resting majestically high on the shipways, looking like a giant torpedo with a sail on the top with two wings. Black as night, she looked ominous and deadly and ready for her role in the Cold War. It was rather romantic, actually, to see this beautiful machine waiting patiently to be built. She looked like she was waiting to be made up for the big dance.

We went aboard Hull 572 by simply walking through a giant hole in her port side, right where the officers' wardroom would later be located. Inside there was raw metal everywhere, mostly unpainted and shiny. Most major components such as the diesel engine, sonar arrays, and supporting interior equipment such as radio antennas, etc., had been installed, but they were not connected and were far from being operable. Access to the equipment was rather easy, and this enabled the crew not only to access the equipment for inspection, but to monitor equipment testing and document any deficiencies identified.

One of the interesting features of Hull 572 that I noticed when walking aboard through the opening in the port side of the hull was the thickness of the steel of the pressure hull. The pressure hull is what protects the crew and the inside of the submarine itself against sea pressure. Such protection must be afforded by very high-strength steel, and Hull 572 was constructed of thick high-strength steel called HY-80, a high-yield steel with minimum yield strength of 80,000 lbs/in^2 or 40 tons per square inch, quite sufficient to withstand the sea pressure at the boat's test depth. The pressure hull must also be nearly perfectly round because even a small deviation such as one inch from the roundness of the pressure hull can cause a reduction of over 30% of designed hydrostatic load. The pressure hull is also strengthened by stiffener rings, and any welding that is done on the pressure hull must be precise and tested by X-Ray to ensure its strength.

With submarine hulls, there is a delicate balance that needs to be achieved between the weight of the steel in the pressure hull and the ability of the submarine to float. For example, someone might suggest that instead of having HY-80 steel two inches thick, why not have it three inches thick? That would certainly add strength to the pressure hull. Sure it would, but then the total weight of the pressure hull would be greater than the water the submarine displaces. The submarine couldn't float at all because it would be too heavy. The only way to keep it on the surface would be to attach a large cable to the hull and hold it on the surface, and when you want it to submerge, let out the cable. But that wouldn't be a submarine . . . it would be a bathysphere! Totally impractical! The total weight of the metal, equipment, and personnel on the submarine must be less than the water it displaces—just ask Archimedes! That Greek got it right!

It's interesting that at around the same time as the construction of the *Ray*, the Soviet Union was constructing pressure hulls for some of their newly designed attack submarines out of titanium—much stronger than steel, and gave the Alpha class submarine an alleged test depth of around 4,000 feet. That was impressive, but I'm not sure if there is really any purpose for submarine operations at such a depth.

Seeing all of the equipment in the forward part of the ship was an enlightening experience for me, because I had been so focused on nuclear power for the past couple of years that I had nearly forgotten that the purpose of the nuclear power plant and, in fact, the submarine itself is to put the equipment in the forward part of the boat in positions anywhere in the world and enable them to "do their thing." The nuclear power plant was important, to be sure, *but that wasn't the purpose of the submarine or even part of its mission.* Without the equipment forward of the reactor plant and the crew required to operate and maintain it, it would make no sense to even have a submarine!

As the Electrical and Reactor Control Officer, my responsibilities in the construction phase were in the nuclear power plant alone—the part of the submarine that pushes the "business end" of the submarine around the ocean. I made a promise to myself that, to the best of my ability, I would try to keep track of what was going on in the construction phase up front, even though I understood that would be difficult.

A couple of weeks after settling into a routine as E & RC Division Officer, both Owen Brown and I were called into Captain Kelln's office. He told us that there was a billet that had just opened up on the *USS Queenfish* (SSN 651), which was a Sturgeon-class (637 class) submarine just like the *Ray*, but was about nine months ahead of the *Ray* in construction. The *Queenfish* was ready to be commissioned and was going to be transiting to Pearl Harbor, Hawaii, its home port. The plan was to put the *Queenfish* into one of the testing programs because, like the Sturgeon being built by Electric Boat in Groton, Connecticut, the first boats of a class are generally used for test purposes.

Captain Kelln said that one of us would be transferred to the *Queenfish* and the other would remain with the pre-commissioning unit of the *Ray*. Obviously the lure of Hawaii and the prospect of immediately receiving sub pay on a commissioned submarine were attractive features, particularly to a married officer like Owen. I pointed these out to him, but I began to sweat a little because I feared that he might have recognized the enormous opportunity offered by the *Ray* and would choose to take my dream away from me.

To my relief, however, Owen came through and selected the *Queenfish*—and I had to "settle" for the *Ray*! Again, I thought about the law of the Navy that presented itself so prominently during the boat selection process at submarine school:

"Say the wise, how may I know their purpose, then acts without wherefore of why.

Stays the fool but one moment to question, and the chance of his life passes by!"

By now I knew absolutely that I was doing "something special," and I celebrated that night by myself with a six-pack of beer and a delicious family-sized pizza with all the trimmings!

The non-nuclear crew and the officers assigned to departments in the forward part of the boat were busy as well. As an example, the *Ray*'s sonar system was the most advanced in its time, being a vastly improved version of the systems that were installed on such boats as *Thresher*, *Flasher*, and other earlier attack submarines. The sonar system serves as the "eyes of the ship" even though its vision was limited to the sounds in the sea, of which there are many. Once you're submerged, sonar is vital to the ship's existence—*if you can't locate your enemy, you can't kill him.*

The sonar equipment was installed in four main locations in the ship. The Sonar Control Room (SCR), where the operators were located, was just aft of the periscope stand on the starboard side of the operations compartment. The Sonar Equipment Space (SES) was on the deck below and held most of the processing "boxes" that did the signal processing.

The sonar sphere was 15-feet in diameter with transducers mounted on it to transmit energy and receive sounds in the water. The sphere was "hard," meaning that it could withstand sea pressure. It had four large preamplifier cabinets inside and was located in the very nose of the ship, the bow—the location farthest from the ship's internal noises. The under-ice sonar was installed in the ship's control room, where it could be under the direct control of the Officer of the Deck (OOD) when operating under the ice in the Arctic.

The transmitting and receiving elements of the various sonar equipment were scattered away from the sonar dome surrounding the spherical array to the very stern near the propeller, including the sail as well as the vertical fin on top of the hull. Understandably, these acoustical sensors seemed to be everywhere. The sonar gang—*one of the best groups of sonarmen ever assembled on a single submarine*—spent a lot of time on the boat as it was perched on the shipway, inspecting the equipment, particularly the transducers mounted on the sonar sphere. Access to the sphere was easy when the boat was not in the water; the only other way to access the sphere would be when the boat would be in a dry dock. Our key sonarmen had been trained on the earlier systems and spent a lot of time going through the books and learning about all the improvements. Several were sent to schools for a few brand-new auxiliary equipment, and a few were sent temporarily to operating submarines to get at-sea experience.

There are hundreds of things that must be done before placing a boat in commission, and some of these have nothing to do with the boat's equipment. For example, a boat must have security watches topside when in port to ensure nobody gets aboard to do any harm. These watch-standers are armed according to where the ship is located and the potential threats to it. Before you arm someone, however, you must make sure he is qualified on the weapon and is trained to use it. The forward crew took care of this by training lectures on the barge with hands-on (but no ammunition) sessions.

Then they scheduled the use of the firing range at the Fort Eustis Army base north of Newport News up the peninsula. We scheduled the firing range training to take place in the mid-afternoon, since once you have sailors well away from the boat it can be hard to get them back. A bus took some of them, but most carpooled to the range. The Army's Range Safety Officer gave the required safety lecture and introduced proper procedures. Then he turned over his job to one of our officers and departed. Most likely he felt he would be safer somewhere other than near sailors firing live ammunition with handguns.

The qualification firing went off without a hitch, but a considerable amount of ammunition was expended. Sailors like firing weapons, so we suspected they chose to miss the targets intentionally until it was time for them to be graded for accuracy. They improved at that point and became qualified. As planned, the firing took the remainder of the work day and the troops disappeared afterward.

Aside from taking a daily walk-through of the boat, our early days as members of the PCU required us to write operating procedures for each system in the nuclear power plant. One shouldn't be naive to think that such operating procedures were already written—*they weren't*. We had to plow through each technical manual for each particular system and write procedures for normal as well as emergency operations. Then we had to go aboard the boat and actually see where these pieces of equipment and related wiring would be located. Captain Ken Carr, who was the commanding officer of the *USS Flasher* in Groton and who took me on my tour of the *Flasher* under construction was certainly right. You DID learn how a nuclear power plant and submarine worked by building them!

I had found an apartment in Newport News not too far from the shipyard and in a nice, quiet area. The apartment was a narrow two-story apartment which was new, clean, and comfortable. Lt. Pete Moffett, an officer from a Polaris submarine, had agreed to share the apartment with me, so I needed to pay only half of the rent, which was a relief to me because my Navy salary was not supplemented by submarine pay. Pete was very tall, about 6'5," and quiet. Shortly after we began to share the apartment, he left to join the crew of his Polaris submarine. As a member of the "Blue" crew, he relieved his "Gold" crew counterpart as weapons officer and prepared to embark on a two-month patrol.

Back at the *Ray*'s barge in the shipyard, we worked frantically on the development of operating procedures over the next month because we knew that on June 21, 1966, Hull 572 would be launched, sliding down the shipways and into the James River . . . then to be towed its berth at a pier very close to our barge. At that time the process of "fitting out" would begin. That meant building, testing, and taking Navy ownership of all the stuff that would be inside, like a complete nuclear power plant, etc. As I thought of this, I began to appreciate the planning that went into submarine construction—*you have to put the big things inside the pressure hull BEFORE you seal it up.*

Other members of the non-nuclear crew—such as those who operated and maintained auxiliary equipment such as the diesel engine, the oxygen generators, the CO_2 scrubbers, the carbon monoxide-hydrocarbon burners, high-pressure air compressors, etc.—were busy as well. Their equipment had to be inspected. Much like the sonar gang who inspected the sonar sphere, the auxiliarymen inspected the boat's external tanks such as the main ballast tanks, oxygen tanks, and the like, which were located outside the pressure hull of the boat.

As the most modern and powerful nuclear attack submarines ever built, the 637-class submarines were equipped with the most sophisticated weapons systems in the world. The major purpose of the submarine was to be the most effective, quiet, and capable anti-submarine weapon yet known to man. Deeper-diving, faster, quieter, and in possession of the largest arsenal of weapons in the free world, plus the most modern sensors and weapons control systems, the *Ray* was the ultimate weapon.

Together with the sonar gang, the fire control technicians were the "brightest of the bright" and worked in tandem to ensure the proper installation and calibration of this new equipment. Much of the work could not be done tied up at the pier and had to wait for actual sea trials where the equipment could be fully checked out and calibrated in order to operate as a fully integrated system.

Imagine the challenges facing the fire control and sonar teams. They would be at sea in a relatively noisy ocean, they would have to detect and classify another submarine, track it, develop a solution regarding its course, speed and range, calculate a solution that could be transmitted to a torpedo, and safely fire and guide that torpedo to its target. To complicate matters, all of this needed to be accomplished even though the target submarine might be nearly silent and hard to detect above the ambient noise level of the ocean.

While the *Ray* never fired a shot in anger, it spent much of its time tracking and classifying other submarines. Both the fire control and sonar gangs worked closely together during the fitting-out period, primarily on getting the equipment ready for operational service. Later on, even during operations, they worked closely together to devise new and innovative ways to use this new equipment, leading ultimately to new techniques that were later available to and adapted by the entire US nuclear attack submarine force. These efforts would eventually result in the *Ray* being the major contributor to the nuclear attack submarine force in tactics and techniques, and earn it the title as the first *"Super Nuke."*

These bright and energetic men did not stop at learning about just their own specialties, a characteristic that made all of them true submariners. While in the shipyard, they took advantage of training opportunities in other areas to broaden their knowledge. As Mac McCoy FT1 (now FTCM [SS] USN Ret) put it, "During new construction, in addition to our own divisional and departmental training schedules, I elected to attend as many of the shipyard-produced training sessions as I could, and came away with a wealth of mechanical knowledge which later on made me a killer of a checkout signer on the new reportees. Some of the things I learned included how the main propulsion shaft was constructed, how and why it was filled with sand, how the flexible coupling and resonance chamber worked, why the bull gear had seven different-sized holes in it, the rubber shock mountings supporting the decks to provide sound isolating, how the scram breakers and the rod raising/lowering mechanisms worked and other seemingly inconsequential things like the concept of sequential-sized breakers. It was really nice to know things like what the Raschig rings in the CO_2 scrubbers did, what hopcolite did in the burners, and other wonderful topics. I look back at the Newport News Shipyard experience as one of the things that really made me a better submariner."

He was not alone. Many of the members of both gangs took advantage of actually building a complete submarine and being in an environment where they could learn about related systems and systems that were important to the submarine, but not necessarily to their own specialties. Again, Captain Ken Carr was right—building a submarine not only helps you learn about your own area of specialty, it makes you overall just a better submariner.

Life at my apartment was generally dull at first until Ann, the woman next door, introduced herself to me when we were doing laundry. She was a divorcée who lived with her three-year-old daughter. She was very attractive, very polite, and we often said "Hello" to one another. On one occasion I invited her to join me for a glass of wine on my back patio, which was adjacent to hers. After that she became a bit needy, generally knocking on the door when I was home and asking to borrow things or if I could go to the grocery store with her.

Then things got a little bizarre.

She started calling me in the middle of the night, telling me that she thought there was a prowler outside. That made me a bit nervous, and when I checked, there was no prowler at all and I hadn't heard any noises. I tried to reassure her, but she wanted me to come over to her apartment and sit with her for a while just to be certain. I did that and sat and talked with her for about an hour until I told her that I had to go back and get some sleep, because I had shift work coming up the next day.

Two nights later the same thing happened and I went over to her apartment and sat and talked with her for another hour. I began to worry a bit about her as well as the security of the apartment complex where we were living. The next day when I came back to the complex, I stopped in to the management office and asked them if there had been any reports about prowlers in the neighborhood. They told me that there had not been any such reports. That night it happened again, and when I sat and talked with her I mentioned that the management office had not received any reports from anyone living in our area.

The next day I talked to one of my shipmates about this and he basically told me that I was "clueless." "There isn't any prowler," he said. "She's just offering you an opportunity."

Clueless was right! I suppose it was my South Dakota naiveté that kept me from figuring out what she was really after, and I felt a little dumb about taking her seriously. When I returned to the apartment that evening, I knocked on her door and spoke to her. I told her that my fiancée was coming for a visit and that it would be rather awkward to receive a knock at the door at two o'clock in the morning from a woman who lived next door. I told her that I would be alert for prowlers and if there was any other way I could help, I would do so.

Apparently the "prowlers" didn't come around again, because she never again knocked on the door again in the middle of the night. In fact, from that time on she seemed rather hostile to me and rarely even acknowledged my existence. (That was fine with me.)

Back in the nuclear propulsion plant, the key word about the watch section was "qualified." In order to qualify us as engineering officers of the watch, we first needed to pass an oral qualification examination conducted by the office of the Commander of the Submarine Force-Atlantic (COMSUBLANT). So one bright and cheery morning, four of us sat down across the table from a senior officer who conducted the examination. The examination was tough, and not all of us passed. Fortunately, I was one who did pass and made it past the first hurdle toward being a regular engineering watch-stander.

After that pre-qualification examination by COMSUBLANT, we were ready to assume watches on the boat: twenty-four hours a day and seven days a week. In order to appreciate what this meant, you have to understand that at that time, the primary threat to the United States was the Soviet Union. The Viet Nam war was important, but decidedly second in priority to the Cold War against the USSR. That was evident in military spending—virtually anything that the submarine force needed, it was provided. And the shipyards were instructed to go on a "war footing" in the construction program—24/7—and we, as the PCO unit that had to supervise the testing program and take delivery of systems as they passed the tests, had to be on board constantly.

But this frenetic testing process had not yet begun, because Hull 572 was still perched high on the shipways, looking like an impatient predator seeking release. The holes in her sides had been sealed, she received a new coat of shiny black paint, and she was ready. As the launch date approached, we were preparing to begin the shift work necessary to stand watches on the submarine as well as conduct the tests along with the shipyard on various systems to ensure that they met all specifications. This meant that a watch section of Navy nuclear-qualified personnel would supervise the testing of each system as it was completed and, if the system met the test satisfactorily, the Navy would "take delivery" of the system. After "taking delivery" of a system, the Navy would be responsible for its maintenance.

During the week before her launch, we attended a few lectures about how the launch would be conducted: the technical details of exactly how the shipyard would transfer the weight of the enormous hull from the keel blocks to the skids on which the boat would slide backwards into the river. To manage the launching of the boat took a bit of careful thought and timing, because once the weight of the boat had been transferred from the keel blocks to the skids, the boat WOULD begin to slide backwards regardless of whether the formal program or speeches had been completed.

On a bright Tuesday morning on June 21, 1966, many events affecting the United States were taking place around the world. In Southeast Asia, US warplanes struck North Vietnamese petroleum-storage facilities in a series of devastating raids as part of Operation Rolling Thunder, which had been launched in March 1965 after President Lyndon B. Johnson ordered a sustained bombing campaign of North Vietnam. In Newport News, Virginia, after a series of short speeches which those of us standing high up on the fairwater planes of the submarine could not hear, we could hear some loud pounding as the wedges were lifting the boat off the keel blocks and onto the skids. Then we felt a slight "lurch" and Hull 572 began the slow but accelerating backward glide down the shipway into the James River. Super Nuke was underway! *Thankfully*, we all were thinking, *it floats!*

Photo Courtesy of Huntington Ingalls Industries

The tug boats were waiting patiently for us, and when Hull 572 stopped moving, they quickly towed the boat over to its pier.

Fitting out the Super Nuke had begun!

Chapter Seven
FITTING OUT

Fitting out a ship is a term that applies to the construction of a ship after it has been floated and before it is delivered to its owners. How that applied to Hull 572 was that we had a shell, not a boat, and it had many of the pieces of equipment installed inside the hull, but not connected or operating. The forward one third of the boat was "non- nuclear" and the aft two thirds of the boat consisted of the nuclear power plant which, unlike the forward third of the boat, fell under the supervision of the Division of Naval Reactors. This chapter will focus on fitting out the aft two thirds of the boat—the nuclear power plant.

Standing watch in the nuclear power program meant having to be "qualified" to stand that particular watch. Much like the prototype, qualification on a nuclear submarine power plant meant standing watches on each particular watch station and getting signed off or qualified on that watch station before moving to the next watch station. Engineering officers of the watch (EOOWs) were required to qualify on every enlisted watch station before being qualified as an EOOW.

Not so with my situation. We were installing a relatively typical S5W nuclear power plant with the General Electric core 3. In actuality, I never stood a training watch on any watch station on an S5W nuclear power plant, because this was new construction.

The process we adopted was new construction "provisional qualification." As systems such as the Reactor Plant Fresh Water system was ready to be tested by us, we would become "provisionally qualified" on that system in order to run the tests on the system. When the tests had been completed and turned over to the Navy, we were "qualified on the system."

That differed from a fully operational nuclear power plant, but doing the qualification process that way at least made me understand the systems far better than I ever would have understood them had I been required to just stand watches on a particular watch station. Again, Commander Ken Carr was right—you do learn how a submarine and nuclear power plant work far better when you build them.

We settled into the watch routine and typically conducted one test on a system per day. The process of testing was very disciplined, and always started by the issuance of a test document—green in color—that had to be reviewed by the upcoming watch section. If the test document applied to a particular system where piping was concerned, then the document was checked against the specific drawings either in the appropriate technical manual or shipyard drawing. In all cases, the test document specified initial conditions, which generally included a specific valve-by-valve lineup, and this lineup would be set up by the Navy watch section, one watch-stander making the valve lineup and another conducting a specific valve-by-valve double check.

The shipyard provided a "Shift Test Engineer" (STE) who stood watch alongside the Engineering Officer of the Watch (EOOW). Whenever tests on particular systems were held before the Navy took possession of those systems, the STE had to give his approval for the test to begin, and the EOOW had to concur in that decision. The tests were conducted in a very methodical and step-by-step process, actually reading each step out loud to ensure that the test was being conducted exactly as written. If a problem with the test occurred during testing, or if an error in the test procedure became evident during the testing process, the test was abandoned and conditions for that system were returned to the beginning set-up situation. To the credit of Newport News Shipbuilding and Drydock Company, errors rarely occurred.

For the testing phase of fitting out, we had three watch sections and the routines became rather similar. Generally we came into the shipyard about four hours before the scheduled watch, spent eight hours on watch, and then spent approximately four hours after the shift ended, for a total of sixteen hours per day on average. Since we were on a war footing, this went on seven days a week with little or no time off for doing anything other than trying to get some sleep. During a typical watch shift—day or night—there would be a wide variety of shipyard workers aboard, each doing his own work. The type of work could be determined by the color of the individual's hard hat and the designation on the front. Individuals who were not a part of either the mechanical or electrical teams wore white hard hats. This included the quality control inspectors and, in particular, the representative from the Division of Naval Reactors (NR)—Admiral Rickover's "spy."

The representative from NR was a nice guy, and we certainly visited with him whenever he would come on board, but most of the time we kept an eye out for him and had one of the members of the watch section follow him around. Behind his back, however, we referred to him as a "White Rat."

We would note anything that he would be looking at closely and, if it seemed a bit out of the ordinary, we would call the captain at home and tell him. It was not that we were paranoid about NR—it was just that Admiral Rickover was known to have his "spies" find something wrong on one of the boats and then Rickover would personally call the commanding officer at home and ask him about it. Although that never happened during construction of the *Ray*, we were always ready for it.

Marital status played a part in watch-standing, but I will admit that on board the *Ray*, it was for the most part understandable and entirely fair. It did turn out, however, that on specific holidays that were family-oriented, I found myself, as the only bachelor in the officer watch sections, having watch for the full twenty-four hours of the holiday. During this time, however, the shipyard was not inactive. Instead, they took advantage of the relatively quiet conditions in the shipyard and, for three days around Christmas in 1966, towed us to a large drydock on the northern part of the shipyard. There, throughout the day and throughout the night, three of us were on board the boat wearing headsets to minimize sound and one watch-stander was on the pier with the shipyard personnel. We would methodically start individual pieces of machinery—mostly pumps—and the shipyard would measure the noise level that the operating piece of equipment emitted.

This was to check the integrity of the sound isolation equipment and structures for every piece of operating equipment in order to ensure that no machine noise was being transmitted directly to the hull and thence to the open water outside the hull. It was painstaking work, but exceedingly important because the absence of any noise emitted by operating machinery meant that the submarine would be very difficult to detect by an adversary's passive or listening sonar.

Shift work and the demands for 24/7 construction left little time for a social life, even for the married officers. With the experience of "Miss Maryland" in the past and a relationship that was only starting to bud in Groton with a Connecticut College student, my social life was asymptotically approaching absolute zero.

Given only eight hours per day available for such luxuries as sleep, going to the cleaners, and attending to the maintenance of my little green Volkswagen, any attempt at a social life was tedious if not impossible. Shortly after the launching, my high school "sweetheart" visited the tidewater area and we had a chance to reignite a flame. That was short-lived, however, because she was headed for central California where she would be a teacher, and trying to combine any relationship while building a submarine, with three thousand miles of separation, was nonsense.

To make matters a bit more complicated and without going into detail, since we were dating in high school, she had become my "stepsister" because her mother had married my father. That alone was a source of sufficient discomfort and "weirdness" to put that relationship in the permafrost for keeps. She later visited during the period between Christmas and New Year's in 1966, but it was clear by then that while we did like one another, anything beyond just a platonic "stepsibling" relationship was not even a remote possibility. The "flame," if it had been there at all, had been extinguished, and it was clear that she did not know what life would be like as a Navy spouse, particularly that of a nuclear submariner. The ending of the relationship was friendly, with no regrets. With that out of the way and finally settled, it was time for 100% concentration on the job at hand: *building the Ray.*

The major project in constructing the nuclear power plant was the completion of the primary system—that is, the nuclear reactor vessel, the main coolant piping, the steam generators, and the related materials that were necessary to function within the reactor compartment. The first big test and procedure with the primary system was the "initial fill," which occurred without having the reactor core installed. The purpose of filling the system with water first was twofold: 1) to check the integrity of the entire system for leaks, weaknesses, etc., and 2) to flush the system and clear it of any residual metal shavings, impurities, and the like.

Heating up the primary system without a nuclear core was

rather simple, but took time. The energy source was the energy put into the primary coolant by the main coolant pumps — a relatively slow process, but heating up a major primary coolant system slowly was essential to ensure that all the metal in the piping system heated at a constant rate, and not at a rate that would cause any change in the metallic structure of the piping itself. This was simple metallurgy at work. Without divulging any confidential information, one could say that the water was heated up to a very high temperature—far above the boiling point of water at atmospheric pressure—and that the pressure maintained in the primary system was sufficiently high so as to ensure that the high-temperature water stayed in a liquid form and did not flash to steam.

When the primary system flushing and hydrostatic testing had been completed, the system was prepared for core load and for the next couple of weeks we focused on various systems in the secondary (steam) plant as they became ready to turn over to the Navy. During this period, the shipyard concentrated on loading the nuclear reactor core—the heart of the nuclear power system—and resealing the reactor pressure vessel. When that was completed, we were, in a sense, loaded and ready to go, but with a lot of equipment testing and plant integration left to accomplish.

I can remember clearly the day of January 15, 1967—not because we were conducting any special nuclear test, but because that was the day of Super Bowl #1 between the Green Bay Packers and the Kansas City Chiefs. Of course I had the EOOW watch and was unable to watch the game, and there were no VCRs or TIVOs back then to record the game. It didn't seem to be such a big deal, because everyone assumed that the traditional National Football League would crush the upstart American Football League—and when we learned that Green Bay had won by the score of 35-10, no one was surprised. We had the discipline, however, not to have any radio play-by-play broadcasting taking place aboard the boat while we were doing the hydrostatic testing. Admiral Rickover would have been very proud of us. We were getting ready to fill the primary system with water with the nuclear reactor core installed. That was going to be exciting.

A few days later I had the watch for the initial fill. This was a potentially tricky step, at least in my own mind, because we were actually adding the "moderator" (water) to the primary loop with the nuclear reactor core. The moderator slows down fast neutrons and enables them to fission with U^{235}, and that got my attention. The reactor was shut down, so there was really no worry at all, but the thought of putting the reactor vessel into position to sustain a chain reaction was a bit exciting. We accomplished the second fill and subsequent primary loop heat-up without a hitch, and the subsequent hydrostatic test of the primary loop with a fully fueled nuclear reactor was successful.

The next couple of weeks were rather mundane, but it became apparent that the former hollow shell that was the beginning of the Super Nuke was slowly filling up with equipment. The pace of construction seemed to be actually accelerating, and quite visible and obvious changes could be seen almost daily when taking a routine walk-through of the forward areas of the boat. Wires and piping seemed to be everywhere, and the best description I can think of for the interior of the boat was that it looked like complicated, but rather organized clutter. *Ray* was beginning to look like a submarine – albeit one in need of substantial clean-up and with some very obvious holes in the top of the hull that needed to be closed and sealed.

The month of January 1967 was a blur, with the pace of construction picking up to near- breakneck speed. I can remember hoping that as much attention was being paid to the forward part of the boat as was being attended to aft of frame 57 (the forward frame of the reactor compartment). Other officers had reported aboard and were focused on the areas of the ship not under the auspices of Admiral Rickover. We now had a weapons officer who supervised the weapons and fire control systems, a sonar officer whose responsibility (obviously) was the sonar division, and a damage control assistant who supervised all the equipment that was non-nuclear—equipment such as the diesel engine, the high-pressure air compressors, the oxygen generators, and the like.

While the pace of work in fitting out the *Ray* was nearly at its peak, the Newport News shipyard was simultaneously building other submarines. For example, the *USS Lapon* (SSN 661) had been launched in December, and sitting up on the shipways still waiting to be launched were the *USS Hammerhead* (SSN 663) and *USS Sea Devil* (SSN 664). There were other ships under construction at the shipyard as well. A few piers up the river, the *USS John F. Kennedy* (CV-67), the last conventionally powered all-purpose aircraft carrier, was being fitted out for service.

The sheer size of that ship ensured that whenever one walked around the shipyard, it couldn't be missed.

Other kinds of ships came and went, mostly undergoing routine or emergency maintenance. For example, as a sometimes bridge player, I was participating in a duplicate bridge event in Newport News and was the partner of the captain of the *Esso Lexington*—a very large oil tanker that had come into the shipyard for maintenance lasting about two weeks. He gave me a tour of his ship—the only oiler that I have ever visited.

One night in January 1967 I had the "mid-watch" on the *Ray* and was conducting some system test that occupied my full attention from the start of the watch at midnight until completion at 0800 hours in the morning. When I went to the boat to go on watch at approximately 2300 hours, the opposite side of the pier where we were moored for fitting out was unoccupied. Our pier was virtually deserted, with the exception of the *Ray*. After standing the watch and being relieved, I climbed out of the boat and, much to my surprise, across the pier just a few feet away loomed the monster passenger liner, the *SS United States*, in all its glory. Apparently the ship was to undergo a few days of routine maintenance.

SS United States is a luxury passenger liner and was the largest ship of its kind built in the United States. It was expressly designed to capture the trans-Atlantic speed record, which it did shortly after going into service in 1952. You can imagine the contrast between that monster and the *Ray*. It was over three times the length of the *Ray* and over ten times the displacement. The *Ray* rested low in the water with the rounded deck scarcely five feet above the water line, while the *SS United States* towered a full one hundred feet above it.

As I was standing in awe, one of the senior shipyard construction supervisors stopped and stared with me. "I'm always impressed with her," he said, showing great respect for that beauty of a ship. "My father actually worked on her when she was being built here in the early 1950s. Want to take a short tour?" he asked.

"Oh, my — yes, indeed," I responded quickly.

He and I boarded the ship, and he took me up to the bridge where the captain and crew would pilot this beauty as she plowed the seas. The bridge was high above the main deck which, by itself, was equally high above the pier. The bridge seemed to be twice as wide as the entire width of the *Ray*, and there was ample space in which to not only operate the ship when underway, but simultaneously host a relatively large cocktail party.

We briefly toured one of the luxurious first class staterooms and I was impressed, knowing that in the past such celebrities as Judy Garland, John Wayne, and Hopalong Cassidy had occupied these rooms. We then toured the first-class ballroom which, with all the furniture removed, seemed gargantuan in size and dwarfed not only the small confined spaces of the *Ray*, but the entire boat herself! The tour was short, but gave me an idea of what it must have been like to go to sea on such a vessel. As I thought about it, I wondered if I would have gotten seasick aboard her, and remembered that during the short tour I did not feel any motion, mainly because the ship was solidly moored to the pier. In any event, I thought that any traveling across the ocean that I might make in the future would be either by commercial jet or by a fully submerged nuclear submarine.

In all of our nuclear power plant testing on the *Ray*, we followed the "letter of the law," even if what was happening made total sense and was in no way dangerous. I can remember one night—on a weekend, I believe—when I was on watch, the primary plant had been cooled down to below the boiling point of water, and we were taking readings on an hourly basis as part of routine watch-standing. One of the readings was for the steam generator sight glass level that showed mechanically (not just through electronic indication) the level of water in the steam generator. The water in the steam generator had to be maintained at a specific level to ensure safety of the system. With no water in the steam generator, for example, there would be no way to remove heat from the primary coolant and thus the reactor. However, the reactor had not yet been taken critical, nor was it a source of heat at all. Nevertheless, the water level in the steam generator kept going down and down until finally it was not visible at the bottom of the sight glass.

This posed absolutely no danger to any piece of equipment at all, but we spotted the NR representative (White Rat) cruising around the plant and to be on the safe side, I called Captain Kelln at home, even though it was in the middle of the night. He knew the water level posed no danger to the plant and actually chuckled that I would call him with such a minor problem, but I was worried that the NR might have alerted Admiral Rickover, who would have then placed a call to Captain Kelln to ask him if he was aware that the water level in the port steam generator was below the level that could be measured by the sight glass. The call from the admiral never happened, and I'm not sure if the NR representative ever saw the sight glass water level. In any event, it really didn't matter at all, but we followed the rules to the letter.

During new construction and during the three "special operations" on which I participated as a watch officer, we never had an "incident"—an event involving the nuclear power plant that has led to significant consequences to people, the environment, or the facility. I attribute this to several things: 1) the excellence of the nuclear power training, 2) the painstaking effort put forth by the shipyard in designing the test program for the power plant, and 3) the time and effort each watch section took before going on watch to review the upcoming tests. This is pretty typical of the naval nuclear program in general. The legacy of Admiral Rickover was excellence and setting the very highest standards of quality, with no exceptions. *That legacy lasts to this day.*

One of the fun duties of placing a ship in commission is picking the "patch," the artwork that will represent the ship. The ship always extends an invitation to the public to submit designs and in *Ray*'s case, the division that had been assigned to work with the shipyard Public Affairs department was the sonar division. Public Affairs put the invitation to submit a design for the boat in the local papers and shipyard newsletters. Very soon thereafter, the submissions began to arrive. Scores of them flooded in, from children's drawings to exceptional designs by professional artists. One submission had a typo in the address on the envelope. If you look at any typewriter keyboard, the "Y" and "T" keys are adjacent. I suppose it was only a matter of time, but we did indeed receive a letter addressed to *"USS Rat,"* resulting in the sonarmen dashing around acting like mice for a few days.

The sonar gang also influenced the selection of *Ray*'s artwork by screening those designs they favored and then giving Captain Kelln the design they wanted to be the winner. It was selected because it was the best. From then till now, this stands for *USS Ray*:

By early February, most of the tests in the various piping systems as well as the electrical systems in the nuclear power plant had been completed successfully and all systems had been turned over to the Navy. The next obvious step in the process was to wake up the slumbering nuclear core.

Luck always seemed to come my way, and as it turned out, the initial criticality watch was when I had the duty. This is the process by which we started up the patiently waiting nuclear reactor core. We prepared carefully for the procedure because the process of taking a nuclear reactor to the point of being critical is very disciplined, and over the course of the next couple of hours, we slowly went through the procedure step by step.

At the appropriate time when we could see that all of the instrumentation indicated that the nuclear reaction was self-sustaining, I was able to announce over the communications system, "The reactor is critical."

The reactor might have been critical, but it was not putting out any power of any significance. After completing the initial criticality tests, we shut down the reactor in preparation for the next series of tests—heating up the primary system with the reactor instead of the main coolant pumps, and then conducting power range testing by putting the reactor into planned transients. Over the course of the next couple of weeks, we took the reactor critical again and conducted several power range tests to ensure that the reactor would be operating as expected. With all of the tests having been conducted successfully, the next step would be the "fast cruise."

Building a nuclear-powered submarine does not consist only of building a nuclear power plant, just as the purpose of a nuclear submarine is not just to enable the Navy to let nuclear power plants plow through the oceans. The real purpose of the nuclear submarine is "up front" where the sensing systems are: the sonars, the weapons control systems, the communications systems, etc. While we were busy writing procedures for and testing the nuclear power plant, others were busy with monitoring the construction of the forward one-third of the boat.

The non-nuclear construction process was carried out beyond the scope of supervision by NR. Simply put, Admiral Rickover cared only for the nuclear power plant and maintaining the high standards of quality for which he was notoriously correct. Supervision at the shipyard was carried out by the supervisor of shipbuilding.

The process of quality inspection was different; we had our own ship's personnel monitoring the quality of construction at every step of the process, and we also had the shipyard's own quality inspection division monitoring on a constant basis. The only thing missing about the non-nuclear parts of the ship was the heavy hand of NR, so the crew itself became the quality control inspectors. This was no problem because, in fact, some of the smartest crew members were "up front"—that is, in the business end of the boat: the sonar gang, the fire control gang, the communications gang, and the navigation gang, to name a few. These were the individuals whose equipment and expertise the nuclear submarine was intended to make mobile—these were the people who would be in the thick of fulfilling the *Ray's* missions.

But first we had to prepare the crew for going to sea—a "Fast cruise."

Chapter Eight
FAST CRUISE

During the evening of Saturday, February 4, 1967, I had dinner with my sister and brother-in-law at the Langley Air Force Base in Hampton, Virginia, just a few miles from my apartment. It had been nearly ten months since I had seen them at the launching of the *Ray*, and we spent the time at dinner chatting about many of the things that had happened during construction of the boat.

Construction of a nuclear submarine follows a meticulous plan. From the time of launching the boat until the process of operating all the systems in simulated conditions involves system installation, system testing, and finally sealing up the pressure hull of the boat so that it can not only go to sea, but actually dive safely without fear of leaks or catastrophic flooding. Think of it another way: you gotta get all the "big stuff" inside the boat before you seal up the pressure hull. Otherwise you have to cut a hole in it later…not a good idea.

The pressure hull of a submarine is a critical part of the boat. While a 637-class submarine like the *Ray* looks on the outside like a big torpedo, generally all you are seeing is the outer shell.

Inside the shell for a large part of the boat is the pressure hull. Between the pressure hull and the outer skin are external tanks, such as those with high-pressure air, the ballast tanks, etc. The sea is relentless in trying to penetrate the pressure hull. If it is successful, it can lead to catastrophic flooding and the loss of the boat. To get an idea of how powerful this pressure is, think about the four 21" torpedo tubes. They penetrate the hull in order that the boat can fire torpedoes, which are stored inside the boat, to the open ocean.

The torpedo tubes have inner and outer doors that prevent the sea from coming inside the boat. The area of a 21" torpedo tube door is approximately 350 square inches. The sea pressure pushing on these doors from the outside when deeply submerged is very high. At major depths, if a torpedo tube door were to fail, the water would come into the ship, much like shooting a 21" column of water over 750 feet in the air. Death for all hands on board would be instantaneous. A one-inch hole in the hull would not be so bad, but it would still shoot a one-inch stream of water over 750 feet in the air.

The Navy has established a very disciplined protocol for testing a submarine before acceptance and commissioning. This protocol consists of several steps.

Dock Trials: The purpose of the dock trials is to demonstrate that major systems and equipment are ready to support fast cruise and sea trials. These trials are ongoing and take place during the fitting-out period, and when completed satisfactorily, the boat is certified for fast cruise.

Fast Cruise: It's called "fast cruise" because the boat is moored "fast" to the pier. It isn't really going anywhere. But the purpose of the fast cruise is to train the crew to safely take the boat to sea. The fast cruise generally takes place within a week of the first at sea trials of the boat. The fast cruise is conducted entirely by the officers and enlisted crew on the boat and must be unhampered by construction, maintenance or repair work, or any shipyard worker or official coming aboard the boat during the cruise. Fast cruise generally lasts for approximately three full days.

Alpha Trials: The Alpha trial is the first underway period, primarily conducted for propulsion plant testing and the initial tightness dive. This trial is to see if the boat can submerge and safely operate submerged, and that all equipment to support submerged operations is working properly. This includes operating at maximum speed, diving and surfacing, shifting from maximum forward speed to "all back emergency," and the like. Admiral Rickover generally rode all nuclear submarines on Alpha trials and conducted nuclear power plant drills. Alpha trials generally were conducted over a two-day period, and boats would always be accompanied by a submarine rescue vessel.

Bravo Trials: The Bravo trial is generally the submarine's second underway period and first dive to test depth. This trial is rather frightening in that it is the first time the submarine is taken to its maximum depth. All areas of the boat are inspected for leaks and valves are cycled to ensure there is no binding. The trash compactor unit is equalized to external sea pressure and inspected for leaks. Each torpedo tube is equalized to external sea pressure, the outer doors opened, and "water slugs" are fired. (That's scary!) The boat is put through steep angles up and down to test equipment, and the boat conducts emergency blow drills from different depths, including test depth.

Charlie Trials: Charlie trials are acoustic trials. These trials, performed by the shipbuilder, are accomplished to determine, under various conditions of operation, the radiated and platform acoustic signatures of the submarine, the controlling noise offenders (including those which are speed-dependent), and whether the submarine meets its underway noise objectives. Acoustic trials performed by Naval Surface Warfare Center (NSWC) are used to establish the ship's baseline signature for normal operating conditions by performing independent measurements under preset conditions of speed, depth, aspect angle, and machinery line-up. Charlie trials generally last about ten to fourteen days.

Acceptance Trials: The final trial, which lasts only two or three days, is the last trial where equipment is checked and rechecked prior to delivery. When these trials have been successfully completed, the boat is ready for delivery to the Navy and for formal commissioning into the naval submarine force.

Together with the shipyard, the *Ray* Precommissioning Unit had taken what was almost a "shell" with holes in its sides and top and converted it into an incredible machine ready to venture out into the deep and often hostile ocean. Here is an unclassified summary of the submarine:

Name:	*SS Ray* (SSN 653)
Ordered:	March 26, 1963
Keel laid:	January 4, 1965
Launched:	June 21, 1966
Class and type:	637 Sturgeon Class Fast Attack Nuclear
Displacement:	3,800 long tons (3,861 t) surfaced
	4,600 long tons (4,674 t) submerged
Length:	292 ft 3 in (89.08 m)
Beam:	31 ft 8 in (9.65 m)
Draft:	28 ft 8 in (8.74 m)
Propulsion:	One S5W nuclear reactor, two steam turbines, one seven-bladed screw
Complement:	107
Armament:	4 × 21-inch (533 mm) torpedo tubes

Before finishing our dinner, we talked about the next step in the process, the fast cruise, which was to begin early Monday morning, February 6, 1967. The fast cruise is a simulated underway period that prepares a crew for the at-sea environment with the new boat. The crew operates the boat twenty-four hours a day as if it were operational, except for one big detail: *It is moored "fast" to the pier.* The fast cruise provides the opportunity not only to test equipment and systems approaching real conditions, but also to get the crew in an operational mindset and depart from the routine of being in new construction and in port where facilities and shore access are readily available. For the period of the fast cruise—in our case three days—everyone lives aboard the boat: eating, sleeping, and standing watch, just as the routine would be at sea. Drills are conducted to simulate getting underway, submerging, emergency drills, testing of the nuclear power plant, standing and relieving the watch, starting up and shutting down the nuclear power plant, and generally getting the feel of what it will be like during sea trials and afterwards.

We finished dinner and my sister and brother-in-law left to go back to their home in Ellicott City, Maryland. I went back to my apartment to get a good night's sleep because I would be having the duty the next night, and at around two o'clock in the morning before the fast cruise, I would have to conduct a reactor start-up in preparation for the upcoming days "at sea" while tied up fast to the pier.

On board during the day before the fast cruise, there was a lot of hustle and bustle, as though we were getting ready for a long deployment. We were going to be simulating being at sea for three days, so we had to ensure that there was enough food aboard for all the meals and that we would not have to interrupt the fast cruise to get something that we had forgotten.

You could tell that it was still winter because we were having a very light but cold rain and the sky was a deep grey color. There were many hoses and cables from the shore to on board the boat through the two open hatches, but these would be removed by the time the fast cruise started. We were also hooked up to shore power, but after we completed the reactor start-up during the night, we would shift to internal power from the operating nuclear power plant and disconnect from shore power.

Everything we would be doing on board for the next three days would be as close to being at sea for real.

There would be no comings or goings to or from the boat during this time—either shipyard workers or the crew. The only way we could look to the outside would be through either of the two periscopes. The only exception to actually being at sea would be that we would be still tied up to the pier.

Bright and early at 0200 hours (two o'clock in the morning) on Monday, February 6, 1967 I went back aft with my watch section and we conducted a normal reactor plant start-up including bringing steam into the secondary system, shifting to internal power and disconnecting shore power. A few hours later we were in "hot standby" and ready to depart on the fast cruise. We stationed the maneuvering watch and simulated getting underway, and after securing from the maneuvering watch we began to conduct drills. The next three days were drill, drill, and more drills. These were mainly fire drills and the sorts of drills where one needed to respond quickly and with almost automatic action to confront the problem being simulated.

Back aft in the nuclear power plant, we conducted emergency drills primarily involving the loss of reactor power. These drills had been incorporated into operating procedures intended to prevent the loss of propulsion power such as apparently resulted in the loss of the Thresher. We knew that we would have to be proficient in these emergency procedures not only to save the boat from disaster at sea but also because we knew that Admiral Rickover, when he accompanied us during our initial sea trials, would be closely watching us implementing the new procedures.

Under normal circumstances, no one either comes aboard or goes ashore during the fast cruise. But during the second day of the fast cruise, a call came in about one of the officers' wives going into labor. It would be Jim Blacksher's first child. Captain Kelln and Bob Harris, the XO, discussed what to do and decided to let him go ashore to be with his wife. Normally even during routine at sea operations, there would be no interruption, but since this was a fast cruise and had no impact on its purpose, Jim was allowed to be with his wife, who gave birth early the next day.

I recall a situation when standing an EOOW watch back aft that confirmed for me how solid the nuclear power training was…at least for me. We were operating the reactor plant at a relatively low power, but were simulating higher-speed operations. In this simulation pumps take a lot of power to run, and are necessary if the reactor power is going to be high, which is the case if you want to take the boat to flank (fastest) speed. While we were definitely not at flank speed (being tied up at the pier), we were simulating high-speed operations.

All of a sudden all sorts of alarms went off—bells, sirens, horns, etc.,—and some dials on the main control panels as well as in other areas of the maneuvering room started moving wildly. That was totally unexpected. Normally an alarm might sound and it would be easy to determine exactly what was the problem and address it immediately, but this was highly unusual. Since the normal procedure in any sort of alarm situation was to put the reactor plant in the safest possible condition, the reactor operator turned to me and made a suggestion as to what we should do to address the problem.

I almost gave the order, but paused and thought for a moment. Somehow I just knew that the cause of the alarm was the loss of a specific power source. It is a common feature on all submarines to have redundant systems—that is, to have a system that can perform one function and another system that can perform the same functions. In a submarine's electrical system, there are two power supplies that power various control panels, motors, etc.

In this particular case, I remembered that power for the starter motors to start the main coolant pumps came from a certain power supply—the one that had been lost. That meant that if we tried to execute what the reactor operator suggested, nothing would happen and the process would result in a reactor scram. Put another way, if we tried to put the reactor plant in a safer condition, we would have scrammed the plant—and that would not have been a good thing at all! I don't remember actually "thinking" this—it was instinctive. Somehow I just knew.

So I replied to the reactor operator, "Negative."

In fact, we didn't have to do anything at all except restore the power system we lost, which, if I remember correctly, was simply to reset a circuit breaker. That's what we did— and all there was to it.

This example is not to demonstrate how smart I was in this situation. Far from it. The example is intended to show the result of the training we received in the nuclear power program. This sort of thing was drilled into our heads; we had to know the systems cold: how they were designed, what they were intended to do, how they worked, their source of power, and the like. When something went wrong, all of these things would flash through your head and how to deal with the problem would quickly become clear. The same can be said for qualification on a submarine, where each individual is expected to understand completely each system on the boat and be able to operate it or deal with it in an emergency.

Consistent with the purpose of the fast cruise, we stood normal six-hour watches before being relieved by the next watch section, meals were served at the appropriate time, the internal lighting was dimmed during the evening, and members of watch stations and the crew spent the bulk of their time operating and testing equipment as though we were actually at sea. At the end of the three-day fast cruise, we were all "drilled out," but felt confident that the exercise had fulfilled its purpose. We did, in fact, have the feeling that we were "at sea," and all of the systems that were operated and tested worked exactly the way they were designed.

The Ray and the crew were now ready for sea trials and an encounter with . . . "The Admiral"!

Chapter Nine

SEA TRIALS

The following Sunday evening, February 12, 1967, I went to the shipyard to get ready to embark on the inaugural sea trial of the *Ray*. This would be the "Alpha trial," whose purpose it was to see if the boat could submerge and operate submerged safely and that all equipment to support submerged operations was working properly.

Alpha Trials

As I walked down the pier and approached the boat, I was impressed. Instead of looking like something that was under construction, it rested silently moored to the pier like a giant predator. The shipyard had given it a fresh coat of black paint and had put the large white numbers "653" on the side of the sail. This boat was ready for action.

I boarded the boat and went down to my stateroom to get a few hours' sleep before getting up promptly at 0200 hours to conduct the reactor start-up in preparation for getting underway. This would be much like the start-up for the fast cruise, except this time the reactor would be powering the boat as it cruised out into the Atlantic Ocean for its first adventure. I would be the EOOW for the maneuvering watch, which meant that I would be overseeing the nuclear power plant during the time it would take us from leaving the pier until we reached the location (about 60 miles out at sea) where we would be conducting the trials.

As I got into my bunk, I couldn't help but think that this Alpha trial cruise would be anything but normal. First, we would be going out in the ocean to find out if this boat could submerge and surface safely. That was exciting, and I was very confident that the *Ray* would emerge from the trials with flying colors. Second, we would have a very special "rider" aboard—none other than the father of the nuclear navy, Admiral Hyman G. Rickover. I began to worry that he might tell me that he never received the letter from me that he asked for during my interview with him almost three years ago.

We all knew that there was a definite protocol for the admiral when he was to ride a boat. First, he needed some "supplies," which included fresh white seedless grapes, a can of SS Pierce sour lemon drops, a khaki uniform, and something with the boat's name and logo. He also required berthing accommodations (the executive officer's stateroom right by the captain's stateroom), someone to give him a haircut (FT1 SS Owen McCoy was given that honor), and we were told that if we were to encounter him in the officers' wardroom, we were to ignore him and go about our business. (I was to experience the admiral in the wardroom later on during the trials.)

We had a photo and a painting of the admiral—one that we did NOT show him, but which Captain Kelln has kept for his rich memories. (You'll have to use your imagination here . . . the mark on the back of the admiral's head is an image of RED LIPS!)

Photo Courtesy of Rear Admiral Albert L. Kelln USN (Ret.)

Promptly at 0200 hours, I went back aft to the maneuvering room in the nuclear power plant with my watch section and we conducted a normal reactor start-up. A few hours later we were in hot standby, on internal power with shore power disconnected, and with the main turbines warmed up. We were ready to depart.

I heard the announcement that the admiral was boarding the boat and knew that shortly we would hear the "Station the maneuvering watch!" command over the 1MC, the boat's main intercom system. When the command came, we were ready to respond to any "bell" —a common term used to signal what we should do with the main engines . . . such as

"Back one third!" . . . which meant that we were backing up the boat to go out in the channel and were departing.

I had thoughts of seasickness, remembering the agonizing experience aboard the destroyer when I was a midshipman at the Naval Academy, as well as feelings of seasickness when I went on the practice cruise aboard the diesel submarine at submarine school. In preparation for this hopefully not-to-be-encountered event, I had put a polyethylene bag in my back pocket as an emergency means of disposing of the results of uncontrolled nausea. That gave me some psychological comfort and I thought it would work.

A 637-class submarine is designed for underwater operations, and the design is quite unlike a surface ship or even a diesel submarine (which is really a surface ship that can submerge). The hull of the 637-class submarine is round, and when the boat is on the surface, it is about 90% submerged. So it generally plows through the water and you get no pitching up and down; all you get is a slow rolling back and forth, even in rather heavy seas.

Photo Courtesy of Huntington Ingalls Industries

Leaving the harbor in Norfolk, Virginia and arriving to the place where we would normally submerge—about sixty miles at sea—takes a little over three hours. I felt fine most of the way, but after being out in the Atlantic and experiencing the rolling back and forth, I began to feel a little woozy. With my silent psychological partner—the poly bag—I felt that I could make it until we submerged.

One of our watch-standers back aft, an electrician, Dennis Harris EM2 (SS), was watching me as I sat on my stool in the maneuvering room. I looked at him a few times and he would give me a wry smile. Harris had a great sense of humor and was a pleasure to work with. But he kept looking at me and finally with a slight chuckle said, "Mr. Jett, you look like you have a belly full of puke!"

Damn! I thought. He was right! Oh, was he right!

That did it for me, and I quickly grabbed my poly bag, stuck my head into it and puked, to the delight of my watch section, who all roared with laughter. It is an unwritten rule on a submarine that says whoever pukes has to clean up his own mess. Fortunately, all of my "product" was safely confined to the poly bag, and I sheepishly bundled up the bag to dispose of it later. In the meantime, I grabbed another poly bag and stuck it in my pocket just to be on the safe side. When we submerged, there was no rolling or motion from the waves and I had no seasickness problem. That would continue throughout my service aboard the *Ray*, and how Captain Kelln handled my seasickness problem was to put me up on the bridge—on top of the sail—and make me the conning officer going in and out of port. (More later about the fun we had on the bridge.)

The first event in the Alpha trial was to submerge the boat. Before any submarine submerges there is a rather routine procedure that is always taken called "Rig for dive." This time obviously was no exception, because it was the *Ray*'s first experience submerging. Rig for dive involves following a specific protocol whereby an experienced and qualified submariner takes a check-off list of many valves in the boat that must be verified as being shut or open, as appropriate. Once completed, this protocol is then followed by a qualified officer. When the rig for dive procedure has been completed, the officer reports to the officer of the deck (OOD), sometimes referred to as the conning officer, that the "ship is rigged for dive and checked by an officer."

When ready to dive, the OOD gives a command to the diving officer—the officer or senior enlisted man—"Diving officer, submerge the ship." The diving officer then supervises the team that is operating the diving panel, and he opens the main vents that allow sea water to flood the main ballast tanks, thus making the boat heavy enough to sink slowly beneath the waves. The initial dive was accomplished successfully and a short while later after checking the boat for watertight integrity the boat surfaced. We didn't go deep during Alpha trials. The deep dives would come on the next set of sea trials—the Bravo trials.

We also put the boat through its paces, taking it to flank speed underwater and testing its ability to go from flank speed to "back emergency," which is what you do if you want to stop the boat quickly. Remember that the *Ray* was quite large—nearly a football field in length—and when submerged had a displacement of about 4,500 tons.

So it took a little time to stop the boat from flank speed, and an interesting factor that had to be considered when doing the emergency reversal was to stay within the limits of the power put on the main propulsion shaft. A nuclear reactor doesn't care how much power you want; it will give all you want when you ask for it. The limiting factor is to not put too much stress on the shaft—too much could cause it to snap. (Now that is power!) It has been reported that one submarine, the *USS Scamp* (SSN 588) did in fact snap the main shaft because of too much power being applied. The shaft snapped external to the pressure hull at the connection to the screw, so no internal damage was caused as a result.

After these required tests had been completed, it was the admiral's turn. We all held our breath, knowing that he would be coming back aft to conduct some reactor plant drills. I was the EOOW and was starting to sweat. We knew that the admiral would occasionally "kill" one of the watch-standers and make someone else take his place to ensure that we had been cross-trained. So we were prepared for this . . . sort of.

When he came into the maneuvering room, he told the reactor operator, "Scram the reactor!" The reactor operator complied and, of course, we heard the associated alarms go off.

Then the admiral looked directly at me with his steely blue eyes, pointed at me and snapped, "You're dead!"

I was ready for that and collapsed on the deck.

The admiral looked at what I had done and snapped, "Get the hell out of here!"

I got the hell out of there quickly. (I was also prepared for the question, "Where the hell is that letter I told you to send me?" but, to my great relief, it never came. I began to think that the admiral might be mortal after all!)

My Engineering Watch Supervisor (EWS) Chief Burnette handled the scram recovery splendidly, and when we were back on line, I took over the EOOW watch as the admiral departed. We then changed watch sections in preparation for his return to conduct a drill with the second watch section. Relieved, I went forward to my stateroom.

A couple of hours later when I was in my stateroom, I decided to go to the wardroom to do some work and grab a cup of coffee before my next watch. When I went into the wardroom, the admiral was sitting at the far end of the table doing some paperwork. He didn't look at me. I ignored him, got my cup of coffee, and sat down at the other end of the wardroom table.

A couple of minutes later, he said to me, "You didn't need to fall on the deck!"

I was taken aback a bit, thought quickly about what I should say, and, as calmly as I could, responded, "Admiral, when you say I'm dead...I'm dead!'

He looked at me for a moment, smiled (he actually SMILED!), and returned to his paperwork. I then began to ignore him again as best I could. Later on I could tell everyone that I made Admiral Rickover smile!

With the required submergence and surfacing tests having been done successfully, and the required speed tests, emergency stop, and Admiral Rickover drills completed, we surfaced the boat and headed back to the Newport News shipyard. Alpha trials had been successful.

The next step in a few days would be the Bravo Trials, where we would be conducting the deep dives and putting the Super Nuke through some "angles and dangles"!

Chapter Ten
SEA TRIALS - BRAVO

There was only good fallout from the Alpha trials. All of the objectives were met and we WERE able to submerge, as well as (thankfully) surface. While you could never really tell if Admiral Rickover was pleased, there appeared to be no indication that he was displeased. I'm sure he managed to read the riot act to Captain Kelln about something, but the captain never shared that with any of us. The bottom line was that we could proceed with the Bravo trials within a few days.

The purpose of the Bravo trials is to put the boat through her paces in terms of her designed depth ability. The design of the boat should allow full and safe operations even at her test depth. Knowing what sea pressure is at that depth and below, and the potential for catastrophic damage that it can cause in case of any equipment or piping failure, gave us all a bit of an uneasy feeling.

I had been on Bravo trials before, with a Polaris submarine, the *USS George C. Marshall* (SSBN 654) in the spring of 1966 while at submarine school in Groton, Connecticut. There were two events that were firmly embedded in my memory:

1) "Angles and Dangles."

These tests were conducted to ensure proper operation of the boat's equipment under steep angles. I remember the command given by the captain while we were cruising rather slowly at a depth of two hundred feet: "Make your depth XXX feet, maximum dive!" This meant that the boat was going to dive down to its test depth at the maximum down angle allowed, and that is STEEP. We all held on to something as the bow of the huge submarine was aimed for the ocean's depths.

As we descended, you could hear the hull creaking and groaning a bit as we came under the tighter grip of the increasing ocean pressure. The pressure hull was under severe compression, but it was designed to withstand that compression and operate safely at its maximum depth.

At test depth the *Marshall* also fired water slugs from each of its four torpedo tubes. When you simulate firing a torpedo to check the operation of the torpedo tubes, you simply flood the torpedo tube, equalize it with the outside sea pressure, open the outer door, and fire the torpedo tube, which ejects water under pressure to the outside. This means that the inner torpedo tube doors as well as the tubes themselves will have to withstand not only the sea pressure at test depth, but also the pressure needed to eject a nineteen-foot-long torpedo from the tube. In the absence of a real torpedo in the tube, what is actually fired out of the tube is water—hence a "water slug."

The captain then gave the command, "Make your depth 200 feet, maximum ascent!" The boat then started upward toward the surface at the maximum allowed up angle—very steep—just as the dive angle was steep. These steep angles ensured that if anything was stored anywhere improperly, it would certainly slide out of its storage space.

2) Emergency Blow from Test Depth

The next test was to descend to test depth and then test the emergency blow system. This was to ensure that even at test depth, the emergency blow system would work properly—unlike that which most likely occurred on the *Thresher*—and that the boat would be able to surface safely.

We went to test depth (but not at the maximum dive angle this time) and as the boat cruised slowly along the captain gave the order, "Emergency blow—fore and aft." With that, the operator at the diving control panel opened the emergency blow valves, which enabled the very high-pressure air to blow into the main ballast tanks and quickly eject the water from those tanks, even against the high sea pressure at test depth. You could feel even that huge submarine lurch when the high-pressure air emptied the main ballast tanks, and the boat immediately started upward toward the surface. The diving officer was careful as the boat ascended to maintain a controlled, but steep up angle as the boat increased speed toward the surface.

You might remember visual images of the submarine in the movie *Hunt for Red October* virtually leaping out of the water as her bow broke the surface. Well, even this behemoth of a submarine surfaced with a very steep angle and stuck its bow out of the water and then plunged back down to a couple hundred feet before slowly returning to the surface. What a ride!

Getting underway for the *Ray's* Bravo trials was similar to getting underway for the Alpha trials, albeit with a little more tension felt by the crew. We were going to go deep this time, and everyone knew the risks.

I had the reactor start-up again starting at 0200 hours, and fully prepared to get underway early that morning. A few hours later, we stationed the maneuvering watch and were underway, and I had my trusty poly bag handy in my back pocket. I was ready for anything that EM2 Harris could toss at me regarding my seasickness.

We headed out to the diving area, which was not so far out as we would normally go, and where the water depth was slightly less than the "crush depth" of the submarine—that is, the absolute maximum depth which the submarine was designed to withstand. This was a bit comforting, although any emergency that would leave us at the bottom of the ocean at crush depth would be a bit unnerving.

We were accompanied by a submarine rescue vessel and established underwater telephone communications with the ship. When we were doing the dive to test depth, the OOD would routinely communicate with the rescue vessel at incremental levels up to and including when we reached test depth.

After reaching test depth, we ascended to a depth of 200 feet to await the next phase of the tests. This was to return to test depth and fire water slugs from each of the four torpedo tubes.

We all were nervous about shooting water slugs from each of the four torpedo tubes, dreading that we might experience the fury of the sea if there were any failures. Thankfully that didn't happen, and that part of the test was soon over. When that nerve- racking test was finished, we returned again to 200 feet, which started to feel rather comfortable even though there was 200 feet of sea water above us.

Next were the "angles and dangles," which were somewhat fun, although still a bit dangerous. The OOD gave the order, "Make your depth XXX feet – maximum dive!" and the ship went into a steep dive down to the depths of the ocean once again to test depth. When we did this, we could hear noises from various pieces of gear sliding around from their storage spaces.

After spending a little time at test depth, the OOD gave the order, "Make your depth 200 feet. Maximum ascent!" And up we went, holding on for dear life and listening to the hull creak and groan.

We descended to test depth again for the emergency blow drill. Again, as we went down the hull protested, but held its own. When we were ready, the OOD announced that we were going to do the emergency blow.

I was expecting a similar experience when the captain gave the order, "Emergency blow —fore and aft!" And I wasn't disappointed.

We headed to the surface a bit faster than my experience on the *George C. Marshall* and I felt that we jumped out of the water in the manner of a Pacific blue marlin that had just been hooked by a deep-sea fisherman's line. As I recall, however, we didn't plunge back down to two hundred feet, although we did go down below the surface a bit before bobbing up like a cork. That was a ride, too!

With a collective sigh of relief, we headed back to the Newport News shipyard confident that the *Ray* was fully ready to withstand the ocean's pressure at deep depths. We had completed the hard stuff and the trials that were the most worrisome, and now looked forward to a longer sea trial—the Charlie trials to test and conduct basic calibration of the boat's extensive and impressive sonar equipment. With that accomplished, we headed back to the Newport News shipyard.

Charlie trials and an unexpected event awaited us.

Chapter Eleven
SEA TRIALS - CHARLIE

With the Bravo trials out of the way, we began making preparations for the upcoming Charlie trials, which were required to determine, under various conditions of operation, the radiated and platform acoustic signatures of the submarine. Testing would also be conducted to detect any controlling noise offenders, such as noisy pumps, and others dependent upon the speed of the submarine. Tests would be conducted under preset conditions of speed, depth, aspect angle, and various machinery lineups.

It was essential for future operations that any noisy equipment or "sound shorts" caused by imperfections in the sound isolation systems be found and corrected. The objective of the submarine was to be extremely quiet even under high speeds, because we would be conducting covert operations where we knew that others would be listening with their sonars to detect us. Being quiet and virtually undetectable was critical, and I cannot overemphasize the importance of the Charlie trials and other sound tests during the shakedown cruises after commissioning.

There is a joke about how to detect US 637-class submarines. You gather your best technicians, put them aboard your most advanced submarine with your most sophisticated sonar, go out in the ocean, be as quiet as you can, and search all around in every direction. You will hear all sorts of noises in the ocean, from whales and dolphins to shrimp. When you come to the one spot where you hear absolutely nothing, you can assume that a US 637-class submarine is nearby, staring you right in the face! *That was the level of quiet we were seeking!*

The Charlie trials were scheduled a few days after the Bravo trials, and a variety of test equipment was loaded aboard and the sonar gang was busy setting up the various sonar systems.

Then it happened!

During a routine test of the boat's diesel engine, which was located in the far forward compartment, the scavenging air blower exploded—fell apart while running. Such a blower is essential for a diesel engine, because it takes air and compresses it for injection into the cylinders of the diesel engine so that diesel combustion can take place. Without a blower, you have no diesel engine; without a diesel engine, you have no backup power source to use in an emergency if there are problems with the nuclear power plant and it is shut down.

The major problem—*and it was a BIG one*—was that the blower and casing were too large to be removed through the watertight doors and hatches in the submarine. The only way to remove and replace the entire unit was to cut a hole in the pressure hull! This was a big problem, to be sure, because that meant cutting through the high-strength steel in the pressure hull, replacing the unit, and then welding the steel that was removed back into the pressure hull. Such a project would require extensive testing and X-raying of the welds to ensure that the strength of the pressure hull was not compromised.

That problem set the Charlie trials back about a week. It caused a personal problem for me because I had been looking forward to the period after the sea trials and commissioning of the boat to take a little long-awaited leave for a few days to recharge my own batteries. There is a "law of the Navy" that says:

> *When a ship that is tired returneth —*
> *With signs of the sea showing plain,*
> *Men put her in dock for a season.*
> *And her speed she reneweth again.*
>
> *So shall ye if per chance ye grow weary,*
> *In the uttermost part of the sea,*
> *Pray for leave for the good of the service.*
> *As much and as oft as need be!*

I needed a rest from all of the shift work we had been doing for the past many months and was looking for a chance to renew my own speed; but that hope was dashed, thanks to the diesel. (But I survived anyway!)

Everyone wanted the ship to get out to sea after this unplanned delay. While the hull was being "buttoned up," representatives from the shipyard, the Navy representatives at the shipyard, Navy laboratory engineers and technicians, and sonar manufacturer technicians came aboard to attach their instrumentation to certain test points in the various sonars. As soon as the hull welds cooled off and had been X-rayed, we prepared to go to sea. We followed the same routine getting underway—my doing the reactor start-up at 0200 hours, pulling away from the shipyard pier a few hours later, and proceeding out to the testing area where we submerged to enable our sonar gang to finally do their magnificent thing. No seasickness to report this time!

After submerging in our designated area, we conducted the tests we could do by ourselves. The first set recorded the signals at the test points in the sonars at 5-knot increments from 5 knots up to flank speed, and was repeated at several depths. This was done to detect any self-noise that might interfere with our detecting other ships and submarines; if we knew about it, we could run at a different speed and/or depth, and we could also plan how to get rid of it.

Sure enough, we found we were emitting a terrible hum during one fast-speed run. Looking at all the various sensors, it appeared to be coming from the sail. We surfaced and looked all through the sail for something that might not be correctly supported, which could oscillate with water flowing around it. But we didn't find anything. The shipyard admitted the noise was present, but had no feel for what would fix it. Finally, we decided to check it out during shakedown cruise and to address it in our Post-Shakedown Availability (PSA)–that period when the ship returns to the shipyard to fix discrepancies found during shakedown.

The second set of tests was a series of internal noise measurements. The shipyard had been given specifications for the level of noise in each space under several conditions, so the observations were checked against those numbers. Most of the boat met specifications in this area, but the Sonar Control Room did not – the ventilation noise was too high. We put that in our report up be checked out and fixed during the post-trial period.

We met a small ship chartered by the shipyard that had a set of hydrophones dangling into the ocean. A "pinger" was attached to this string. We would approach the pinger at the proper speed and depth for the measurement to be done, but try to just barely miss the pinger and hydrophones. We could usually pass the hydrophone string within 100 yards, giving adequate signals at the recorders on the small ship for post-trial analysis by the Naval Ships R&D Center radiated noise specialists. We were told by the small ship that the "hum" was present but otherwise we looked good, and we received a more detailed report later.

Personally, I stood my required EOOW watches as all the action was taking place forward, and when off duty decided to work on my submarine qualification. Yes, I was a qualified engineering officer of the watch for the nuclear propulsion plant, but I wasn't "qualified in submarines." This meant that I would not be able to independently stand Officer of the Deck (OOD) watches, or Conning Officer watches. I knew virtually every nut, bolt, wire, dial, switch, and lever back aft in the nuclear power plant, but forward of the reactor compartment I was left with little more than what I had learned in submarine school, and I had forgotten a lot of that during the year since leaving sub school for new construction of the *Ray*. I had a lot to learn, and decided that during Charlie trials would be as good a time as ever to start the process. I needed to become "qualified in submarines" and earn my Gold Dolphins—the symbol proudly worn by every officer qualified in submarines.

Qualification in submarines for an officer is very similar to enlisted submarine qualification. It means that not only does the officer have a basic level of knowledge of all of the systems on board the boat; he is capable of performing whatever damage control might be needed under emergency situations. Additionally, the officer qualification goes beyond that of enlisted personnel because of the requirement for the officer to be able to serve as an officer of the deck and drive the boat both at sea, during a maneuvering watch when the boat is either coming or going from its pier, diving and surfacing the boat, and standing watches on the bridge when the submarine is on the surface.

Of particular importance is being qualified to stand Officer of the Deck watches in port— essentially supervising all aspects of the boat when other officers are absent.

With all the activity going on in the control room with the sonar trials, I decided to focus on the auxiliary equipment forward of the reactor compartment. The senior auxiliary petty officer, EN1 Smith (Smitty) was the perfect submariner to teach me: a forward "nup" (non-useful person, in submarine terms) about the operating auxiliary equipment.

Smitty was a jovial man, friendly, and tolerant of an unqualified officer. He was quite willing to help me and "make me useful" up forward. So during the Charlie trials, we focused on five pieces of equipment.

First was the diesel engine up in the bow compartment. Nuclear submarines have diesel engines as an emergency backup to the nuclear power plant. These diesel engines run a generator that provides electrical power to charge the boat's main battery which, in turn, is able to run the emergency propulsion motor in case of disability of the nuclear power plant. The diesel had just been repaired in the shipyard because of the blower motor failure, so we spent time talking about the operation of the diesel, but not actually running it, because we were submerged and there were important acoustic trials going on. After a very thorough explanation while actually looking at the diesel, Smitty gave me a technical manual and told me to read it. I complied, spending time in the wardroom between watches back aft and over several cups of coffee.

We then turned our attention to the boat's atmospheric control equipment.

Bear in mind that when a nuclear submarine deploys on a long mission, it stays submerged for months at a time, limited only by the amount of food that it can carry. An important question to ask is "How do they breathe on board a submerged submarine for so long?"

The answer is not from stored oxygen in tanks—although the *Ray* did have oxygen storage tanks which, without replenishment, could be used in an emergency. The answer is that a nuclear submarine makes its own air from sea water. Sea water is distilled, purified, and then run through an oxygen generator which, through electrolysis, breaks the water (H_2O) into its parts, oxygen and hydrogen. The hydrogen is bled overboard and the oxygen is stored and used.

The *Ray* and other 637-class submarines had two oxygen generators. These were sophisticated pieces of equipment that needed close monitoring because of the danger of hydrogen gas—the unneeded byproduct of electrolysis. Smitty gave me a thorough indoctrination of the oxygen generators and, if I remember correctly, we even started one up and shut it down. Then, as with the diesel engine, he gave me a technical manual, which I took back to the wardroom and read.

There is an officer's submarine qualification notebook that requires you to answer questions in writing. During the course of my indoctrinations with Smitty, I filled out the appropriate areas of the notebook pertinent to the equipment he had explained to me.

Obviously, a submerged submarine is a closed space, without ventilation from the outside. When 120 men are confined in such a space and breathe, they exhale carbon dioxide (CO_2) into the air. Something has to be done to get rid of the carbon dioxide. This is accomplished by equipment known (appropriately) as carbon dioxide scrubbers. They literally "scrub" the carbon dioxide out of the air in order to keep within acceptable limits. The scrubbers used an absorption process with the chemical monoethanolamine (MEA for short) and were highly effective in keeping the carbon dioxide levels low. After cycling one of the scrubbers and understanding how it worked, I took the technical manual back to the wardroom for further study. (By now you should get the feeling that technical manuals come in useful on nuclear submarines.)

There is always the problem of carbon monoxide (which is deadly poisonous) and hydrocarbons that build up in a confined spaces. If you think about confining 120 men in a submarine for one day, the hydrocarbon level that would build up with only one fart from each man would be significant! Carbon monoxide comes from various sources, and one source was from tobacco which, in those days, was allowed on the submarine.

Carbon monoxide and hydrocarbons are literally "burned" out of the atmosphere. The CO/hydrocarbon burners draw pre-heated air through a catalyst bed at a temperature of about six hundred degrees—burning the CO and hydrocarbons—and then cool through a bed of lithium carbonate (Li_2CO_3) that removes any acidic gases. These burners are very efficient.

To remove smoke particles and any dust particles in the air, electrostatic precipitators were used.

The levels of oxygen, nitrogen, and carbon dioxide were carefully monitored in the submarine, and the net result of these pieces of equipment and associated technology on the submarine environment was to create an atmosphere that was actually better than the fresh outside air. In fact, when surfacing the boat after being at sea for several months and ventilating the ship sixty miles out in the ocean, the "fresh ocean air" smells rather fishy. When going into port, the smell of the city is rather putrid, but one's olfactory system gets used to it after a few days and just doesn't notice it. Submariners who had served on diesel submarines certainly knew the difference between the atmosphere on a nuclear submarine and that on a diesel. The diesel smell, once it gets into your clothing, never seems to come out.

After about one week of Charlie trials, we were finally completed with all of the trials where significant testing was taking place. We returned to the Newport News shipyard knowing that only a very short series of Combined Acceptance Trials (CAT) stood between us and commissioning of the boat. After some repairs in the shipyard, we went out to sea again for a couple of days on the CAT trials which, compared to the Alpha, Bravo, and Charlie trials, seemed like a joy ride. The CAT trials might be described as being like taking a fine car with no problems on a short test drive. If you like it, you buy it. We took the *Ray* out on a short test drive, the Navy liked it, and bought it. The *Ray* was delivered to the fleet! Now it was time to make things formal and commission the boat as the *USS Ray* (SSN 653).

My submarine pay was drawing near!

Chapter Twelve
WELCOME TO THE FLEET

The *USS Ray* (SSN 653) was now officially owned by the US Navy! It looked like a fighting ship: black, sinister, lying low in the water, ready to embark on its missions and create new adventures.

Between the time of combined acceptance trials and commissioning, there was work to be done: repairs, some minor testing, tweaking of equipment, and general housekeeping to ensure that the boat would be a suitable place for the officers and crew to live and work during the upcoming shakedown cruise as well as future missions.

After commissioning, we were going to be assigned to Submarine Squadron Six based in Norfolk, Virginia and a part of Division 62. That division had two fast-attack submarines: the *USS Skipjack* (SSN 585) and the *USS Scorpion* (SSN 589). We would be moving from the Newport News shipyard over to Pier 22 at the Norfolk naval base alongside a lot of surface warships and a submarine tender—the *Orion*. The tender is a repair ship for submarines and is essentially a floating shipyard that can make many repairs that do not require going to a shipyard.

The commissioning ceremony, which would be held at the Newport News shipyard, was about a week and a half away, scheduled for April 12, 1967. This was two days short of a full year since I had reported to the *Ray* when it was just a hulking shell perched high on the shipways. Much work had been done since then, and much had been accomplished. My thoughts went back to the advice given me by Commander Ken Carr when I was in submarine school: the way to really learn about a submarine is to build one. Those were very wise and accurate words.

My USNA classmate Frank Spangenberg had recently been assigned to the *Ray*, and since we both needed to find a place to live in Norfolk, we decided to search for an apartment, and room together. Norfolk is quite a military town and thus had many apartment complexes to choose from, given the high turnover rate of military personnel coming and going. We picked a nice apartment at "Pinewood Gardens," not far from the naval base, and a place where other submarine officers and their wives were living. Specifically, we were neighbors of two USNA graduates—one classmate and one from the class of 1965 (the year after our graduation). Both were officers assigned to the *USS Scorpion* (SSN 589). Little did we know what tragedy was in store for them and for their families a little over a year later.

Commissioning of the *Ray* took place on Wednesday, April 12, 1967 at the Newport News shipyard. It was a formal affair, with the shipyard and Navy dignitaries along with the boat's sponsor, Mrs. Thomas (Betty) Kuchel as well as Senator Thomas Kuchel. The officers and crew were all decked out in their finest uniforms, and with little fanfare, Hull 572 officially became the *USS Ray* (SSN 653).

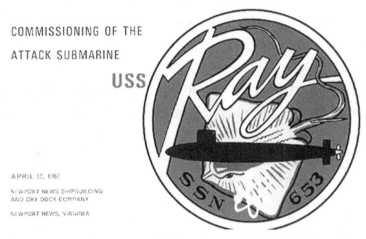

Document Courtesy of Huntington Ingalls Industries

The day after the commissioning ceremony and related receptions and parties, we got underway briefly for the short voyage across the mammoth Norfolk harbor among the various naval surface ships and aircraft carriers. We tied up to our new home, Pier 22, at the Norfolk Naval base.

The next couple of weeks were spent getting the boat ready for shakedown cruise. This meant that the boat would be going to sea at various intervals and testing equipment under real working conditions, as well as enabling the crew to become familiarized with the boat as an operating ship. Major testing conducted during the shakedown cruises was focused primarily on the weapons and sonar systems to ensure proper calibration, that the systems would work together as designed, and that all weapons systems were properly aligned. Ultimately, when the testing was completed, the operating crew could be confident that they could track any contact accurately and, if necessary, employ the appropriate weapons. At the end of the shakedown cruises, the ship would be going back into the shipyard at Newport News to repair or replace any equipment that might turn out to have problems or did not work properly. Following this post-shakedown availability, the boat would be ready for deployments—to do what it was designed to do.

These shakedown tests were conducted in Roosevelt Roads, Puerto Rico, the deep water off the western coast of St. Croix in the US Virgin Islands, and in the Tongue of the Ocean (TOTO) located in the Bahamas. These visits enabled the crew to enjoy some well-earned liberty and time ashore in fun places, and they took advantage of those opportunities.

The water depth off the west coast of the island of St. Croix in the US Virgin Islands drops off very sharply into an abyss several thousand feet deep. This creates an excellent place to conduct sonar tests—which we did—but also afforded us some fun. Since the water was quite deep even near the shore line, we could approach the island and the town of Fredericksted submerged and be safely at periscope depth just a few hundred yards or so off the main pier at the town.

As we approached the pier at periscope depth, you could see along the beach to the north of the pier many tourists enjoying the sun, swimming, and beach life in general. We surfaced the ship and it was rather amusing to see the reaction of the people on the beach when seeing a huge black monster silently appear from the depths of the ocean right before their eyes! There was a lot of excitement on the beach, with people pointing at us.

We proceeded slowly over to the long pier that jutted out into the water from Fredericksted and tied up for the evening. Unfortunately, the pier was not equipped with shore power, which meant that we would have to remain online throughout our stay overnight. On a personal level, that meant that as the only qualified bachelor on board, I would have the on-board duty and would spend my evening in Fredericksted monitoring the nuclear power plant. The crew had fun; my watch section and I kept the boat fully powered throughout the night.

The time spent in Roosevelt Roads, Puerto Rico was all about the weapons systems. All aspects of the weapons systems were thoroughly checked: alignment, ability to load and unload various kinds of weapons, moving weapons around in the torpedo room, checking the signals from the boat's fire control systems to the weapons, the input from the ship's inertial navigation system to the fire control system, and the like. The *Ray* had been chosen as the test model for the 637 class submarines for weapons systems, so the tests were long and extensive.

We conducted another cruise to the Bahamas for sonar testing in the Tongue of the Ocean (TOTO). TOTO is the name of the deep oceanic trench in the Bahamas that separates the islands of Andros and New Providence. It is ideal for conducting submarine sonar tests because its location is relatively isolated from the ambient noise of the open Atlantic. These tests consisted of calibration of the sonar equipment, accuracy measurements of both passive and active sonar, and further noise monitoring tests under different conditions of depth, speed, equipment lineup, and the like.

During the course of the week-long testing at TOTO, we arranged for groups to be picked up by a boat at sea and taken into the town of Nassau for a day and night of rest and relaxation. A fellow officer, Jim Blacksher, and I drew the first day of R & R and were eager to go ashore and perhaps do a little gambling in one of the casinos as the boat picked us up to take us into town. We checked in at our hotel around 1600 hours and went to our large room, where there were two large comfortable-looking beds in a room with a nice balcony and terrific air conditioning. We were a little tired, so each of us decided to take a short nap before going downstairs for a nice dinner and perhaps an evening of cocktails and a little harmless gambling.

Those beds were very comfortable, the room was quiet, dark, and cool, and neither of us had any trouble falling asleep. When I awoke later, I felt quite relaxed and started to think about dinner and the evening of gambling. After checking my watch, however, I found that it was 0400 hours—four o'clock in the morning. Jim and I had completely slept through the night, wasting our opportunity for good food at the restaurant, fine rum drinks, and riches at the gambling tables! About all we could do was laugh—which we did —and go back to bed, but first we put in a wake-up call so we wouldn't miss breakfast. Later that morning, we enjoyed a nice Bahamas breakfast and then went to the lodge on the shore where the boat from the *Ray* would be picking us up at around 1600 hours later that day. After a great day in the sunshine by a nice pool, we met the charter boat which took us back out to the *Ray*. Neither of us told anyone aboard about our aborted "adventure."

Following the shakedown cruises, we spent approximately one week in the Newport News shipyard, making repairs to some of the equipment that was found to be a bit noisy during the acoustic trials, and also fixing a vent system inside the boat's sail that was a source of a humming noise detected during the Charlie trials. Following all of the maintenance and repairs, we returned to Pier 22 in Norfolk to prepare for our first adventure: a "Special Operation." The *Ray* had been tested, shakedown cruises completed, all systems calibrated and checked and checked again. The only thing we needed to do was load weapons and food, and we were ready to meet whatever challenge might face us wherever it might be in the world. The *Ray* was moored beside other submarines and was lying low in the water, black and ominous, looking sinister, hungry, and ready to go!

My roommate Frank Spangenberg and I had new jobs on board. Frank had relieved me of my job as Electrical and Reactor Control Division officer (E & RC), and I had become Communications Officer—effectively shifting my responsibilities to "up front" where the action would be happening on the upcoming mission. That was exactly what I had wanted ever since submarine school. I would be in the middle of what we would be doing.

I was working on my submarine qualifications both during my routine days on board and during my days as Officer of the Deck. Both Frank and I had an opportunity to check out the social scene in Norfolk and, much to my surprise, found—just as was the case in Annapolis—a lot of North and South Carolina schoolteachers working in the area and, in particular, living the good life down in Virginia Beach, which was just a short drive away.

We checked out the various officers' clubs and found a winner, the officers' club at Little Creek! What was so fun about this place was the sing-along three nights every week, featuring Pappy Walsh on the piano. Pappy Walsh was a retired Navy musician who played a mean piano. In the basement of the club there was a large room with an upright piano in the back left corner. The beer flowed freely and Pappy would lead the crowd in a variety of songs. Everyone pitched in—and while it didn't get rowdy, it was very festive.

There were many favorite songs that everyone looked forward to singing at each sing-along. As *Virginia Pilot* reporter Dawson Mills reported after Pappy died, "There were old standbys like 'Carolina in the Morning,' 'Walking My Baby Back Home,' and 'I Wonder Who's Kissing Her Now.' There were college songs and football songs and ('all rise, please') 'God Bless America.' And it wouldn't be a sing-along without a rousing chorus of 'The Virgin Sturgeon' or 'Let Her Sleep Under the Bar.' There were, of course, the service songs, starting with 'Anchors Aweigh.' Invariably, the Air Force anthem began, 'Off we go, into the wild blue yonder . . . crash!' before Pappy got around to playing it straight."

It was a great place to go—a wonderful place to meet the schoolteachers—and, looking back, Pappy became a legend. He apparently continued with that regular gig for years after

I left the Norfolk area, and died in 1997 at the ripe old age of eighty-five. Everyone loved Pappy, and he loved the sing-alongs!

As the fall of 1967 approached, we began preparations for our first mission. From my own role as Communications Officer, this involved ensuring that we had up to date cryptographic equipment and intelligence documents safely locked away. All of the officers obtained the security clearances necessary for such missions and we began regular visits to the headquarters of the submarine force (Atlantic fleet) to read previous patrol reports from submarines that had conducted such missions.

This was, in effect, the extent of our training, and at the time, it seemed odd to me that we didn't receive any formal briefings about the nature of the missions, recent activity, and the kinds of situations with which we might be confronted during the course of our mission. We had to discover those things for ourselves from the patrol reports, and we did not have any underway training that might have prepared us for this kind of activity. This was something that came back to me in a big way later on after I was transferred from the *Ray* to the US Submarine School in Groton, Connecticut as a member of the Prospective Commanding Officer training unit.

At the end of September, 1997, we were fully loaded—weapons, food, enthusiasm, and intelligence—and we were ready to deploy to what for many of us would be "uncharted waters."

The time for the Super Nuke had arrived!

Chapter Thirteen
THE FIRST MISSION

October 1967 had its share of notable events, including the Soviet Union performing two major nuclear tests, Thurgood Marshall being sworn in as the first black man to serve on the Supreme Court, the capturing and execution of guerrilla leader Che Guevara in Bolivia, Charlie Finley's success in moving the Kansas City Athletics to Oakland where he would eventually win three World Series trophies and the conspicuous absence in twenty years of the Giants, Yankees, or Dodgers in the World Series. Later in the 1980s, Charlie Finley became a good friend and we had many lunches and dinners together. I had the honor of taking him to "opening night" at Wrigley Field on 8/8/88! He was quite a character!

Preparing for a nuclear submarine deployment that might last a few months was a first for me, but it was what I had been looking forward to since early in nuclear power school. This was it! We would be leaving the next day—it was time to pack for the trip. But what to pack?

Space is very limited on a nuclear submarine, even for an officer. On the 637-class submarine, officers live in small staterooms that are about eight feet by six feet—typically three officers to a stateroom, where there is a locker for each officer, one air-conditioned bunk for each officer, and two built-in desks to share. In order to reduce the amount of clothing needed for even a long mission, each member of the crew had two pairs of coveralls called "poopie suits." These comfortable coveralls, blue in color, were of just one piece and made of polyester in order to reduce the amount of lint in the controlled atmosphere. We would typically bring along a half dozen pairs of boxer shorts (or briefs) and an equal number of T-shirts and socks. Recommended shoes were any kind that had non-slip rubber soles in order to reduce the amount of noise. My selection was the very comfortable Hush Puppies®.

Other than a toothbrush, toothpaste, and some deodorant, there was little need for toiletries. Many would avoid bringing along any shaving gear because they would grow (or attempt to grow) beards during the mission. That was my choice.

If you wanted to smoke, you brought along enough cigarettes to last for the trip—or, in my case, a couple dozen cigars. Smoking was allowed on board submarines those days because the atmosphere control system (the electrostatic precipitators and the CO/hydrocarbon burners) was sufficient to remove odors throughout the boat.

The food on board the submarine for an extended patrol was excellent. Submarine food is often regarded as the best food in the military simply because of the limited storage space in the ship's freezers. Carrying food that had lesser nutritional value—such as fatty steaks, etc.—was not a practice. Instead, the meat carried consisted of lean filets, lean pork, etc. Food items that would be missed during a long patrol included such things as fresh milk and fresh vegetables, particularly lettuce.

Security about the mission was at the highest level. When leaving on missions of this kind, all you could tell even your closest loved ones—such as your wife, kids, parents, current and former girlfriends, etc.—was that you were "going to sea to operate." That was all that could be said, for security reasons.

Special preparations were made for the boat prior to deployment. The boat was triple-checked for any loose gear or for gear/equipment that might cause a sound short (an accidental connection between a source of noise and the hull). The hull was given a fresh coat of dull black paint and all the hull numbers/markings of any kind were painted over. All power connections to equipment that could transmit acoustic signals, such as active sonar, were disconnected. After submerging, all power connections to equipment that could transmit electronic signals, such as communications and radar equipment, were disconnected. Every effort was made to ensure that no noise or signal of any kind would be emitted from the boat during the mission, except in deliberate and very rare circumstances. The reason for this emphasis on complete acoustic and electronic silence was to avoid detection at all costs.

We did not even carry a flag to fly when transiting to and from the diving area. Anyone seeing this boat leaving the harbor would see only a completely black shape sneaking silently out of port to places unknown.

On the day we were to depart, I again had the start-up watch back aft beginning at 0200 hours. A few hours later, we were in hot standby with the main turbines warmed up and disconnected from shore power. The Super Nuke was now free, independent, and ready for its first adventure.

We left Pier 22 early in the morning and slipped quietly out of the Norfolk harbor, into the channel, and out into the Atlantic. Almost immediately we shifted the on-board clocks to Greenwich Mean Time (GMT), which is the local time at the longitudinal meridian that runs through Greenwich, England. GMT is used as a standard for all Navy messages. About three hours after departing, we had rigged the ship for dive and submerged, not to surface until about two months later in the same location after our mission was accomplished. As is the practice with all naval ships, we filed what is called a "movement report," which usually charts out exactly where a ship would be going. Our movement report, however, consisted of just two words: "Mod out." That meant we were proceeding out to sea to conduct independent operations in accordance with our orders and, in general, we just said, "Bye-bye!"

Photo Courtesy of Huntington Ingalls Industries

With that, we changed course and headed north.

We settled in at a comfortable cruising depth and speed and began the long trek north. During the course of the transit, we held many drills to keep the crew and officers sharp and set the watch rotation to establish a good routine. We were not required to maintain any radio contact with the exception of copying one message every twenty-four hours. We did not transmit any messages. After a couple days of transiting, we were entertained by our first event—crossing from the Labrador Current into the Gulf Stream Current. The Labrador Current flows from the Arctic Ocean south along the coast of Labrador, southward around Newfoundland, and continuing south along the east coast of Nova Scotia. The Gulf Stream begins near the southern tip of

Florida and flows northward along the eastern coastline of the United States and north toward Newfoundland before crossing the Atlantic Ocean.

Each is a powerful current, with the Labrador current cold and flowing south and the Gulf stream warm and flowing north.

The two currents meet off the eastern coast of Newfoundland at the Grand Banks, causing a mixing of the cold and warm waters, lifting nutrients from the bottom and creating one of the best fishing areas in the world. The clash of the two currents also creates one of the foggiest places on earth. That mattered little to us because we were considerably south of the Grand Banks, and fog doesn't bother a submerged submarine.

South of the Grand Banks, the two currents flow in opposite directions and are very close to one another, causing a rather sudden change in water temperature as a submarine transits from one current to the other. Since we were transiting eastward, we went from the colder Labrador Current into the warmer Gulf Stream. We felt the effects of the transition in two ways.

First, cold water is denser than warm water, so when the submarine passed from the colder water into the warmer water, it passed from dense water into less-dense water. If you remember Archimedes Principle, this means that the amount of water displaced by the submarine weighed less in the warm water than it did in the cold water, causing the boat to become heavier in the warmer water. While this was not too noticeable at our transit speed, we did, nevertheless, have to compensate for the warmer water by pumping out more sea water from the trim tanks to make the boat lighter and to bring it back to neutral buoyancy.

The second and more dramatic effect was felt by the nuclear power plant, and we were ready for it. The cause of the effect was the rapid rise in "injection temperature"—the temperature of the sea water that enters the main condensers of the power plant. Steam condensers operate with greater efficiency when the injection temperature is cold, simply because colder water is able to absorb more heat, thus more easily converting the steam to water. When the injection temperature is higher, the condensers are not so efficient, and the resulting water (feedwater) from the condensed steam is warmer. This warmer feedwater is pumped back to the steam generators and adds energy to the steam generator, resulting in the temperature of the main coolant flowing back to the reactor to be warmer and less dense than before.

The less-dense water doesn't slow down neutrons as well as the denser water (there are fewer red billiard balls to slow down the white balls), and so the power of the reactor drops, lowering the temperature of the coolant leaving the reactor. This means that the average temperature between the outgoing coolant and incoming coolant is lower. Since the pressurized water reactor is designed to operate at a certain average temperature, the reactor operator has to compensate for this lowering in average temperature by pulling out the control rods a bit to increase the neutron population, bringing the reactor power back up to the required operating level.

All of this is quite simple, but when the warmer sea water is encountered, it sets off all sorts of alarms in the engine room—bells, whistles, horns, etc.–which can alarm those who don't understand what is taking place. All of the watch-standers, however, are trained for this sort of event, and it turns out to be just a rather amusing and entertaining spectacle. In all, the whole transition takes place over a few minutes and relieves the boredom of the normal steady operation of a well-designed pressurized water nuclear power plant. On the *Ray*, those who had never experienced the transition before enjoyed it the most—"nuclear entertainment," if you please!

When transiting in the opposite direction, from the warm water to the cold water, the same sort of event occurs, but the result is that the reactor operator has to insert the control rods a bit to bring down the average coolant temperature to its normal operating level. Reactor operators also have to adjust the control rod height because of the buildup of fission products, called "poisons." The effect of these poisons shows up during extended deployments where the nuclear power plant is in operation over a long period. When the nuclear fuel U235 fissions, fission products are produced. One of the more common fission products is Iodine 135 (I135) which, by itself, does not affect the fission process. However, I135 has a relatively short half-life, meaning that it decays into some other element— specifically Xenon 135 (Xe135) which loves to absorb thermal neutrons. When the level of Xe135 starts to build up, it robs the U235 atoms of available thermal neutrons (Xe135 absorbs the neutrons) and thus has the effect of reducing reactor power, making the average coolant temperature decrease. To compensate for this normal event, the reactor operator has to withdraw the control rods slightly to increase the neutron population and maintain an average temperature. Over time, the rate at which the Xe135 is produced and the rate at which it decays become balanced and there is no more need for control rod withdrawal to maintain the average temperature.

So during the transit, back aft we were enjoying the effect of moving from the Labrador Current into the Gulf Stream as well as keeping our eyes on the average coolant temperature. This was simply normal procedure in operating the nuclear power plant. Over time, the reactor reached a steady state condition and virtually no adjustments needed to be made.

Up in the forward part of the boat, things were a little different. It's a long way from Norfolk, Virginia to the Norwegian Sea, and the transit, at a steady cruising speed of from 16-18 knots, would take a few days. We were all settling into a comfortable routine, and the cooks were busy providing some of the legendary excellent submarine food for the crew. In the Navy, ships have a reputation for being a "feeder" or a "fighter"—meaning that some ships have good cooks and therefore are "feeders," and some ships don't really have good cooks and are, therefore, "fighters." The *Ray* turned out to be a combination of "feeder" and "fighter"— we had good food and we operated a lot.

As we expected, it didn't take us long to run out of fresh milk and some fresh vegetables such as lettuce, and we learned to compensate. Personally, I decided to go on a diet to last as long as the mission. It took a bit of discipline to get on the diet and stay on it, but I managed.

The fire control and sonar technicians were constantly running checks on their systems, because soon the focus would shift dramatically from routine transit conditions. As Communications Officer, there was not much going on because we weren't communicating with anyone; but on a daily basis, we would come close to the surface and copy a routine message or two intended for us. These messages were generally very short and would eventually contain any information about possible contacts in which we would have an interest. Thus far, however, in the few days we had been transiting, there was nothing of interest.

There was not even any news about what else was going on in the world. Since we would be near the surface only long enough to copy messages intended for us, there was no time to locate and copy a news broadcast of any kind. So regarding what else might have been going on in the world, we were virtually clueless. But that didn't seem to matter, except to some who were avid college football fans—they missed the results of the weekly clashes between their favorite teams and their opponents.

Then it happened.

Soon after entering the north Atlantic, we received a message that there was a contact that had been detected on a southwesterly course, headed our way. The message tentatively classified the contact as a submarine and gave us an approximate location as well as an estimated course and speed. That was enough to get our attention, and from the details of the message, we plotted a course and speed for an intercept. There was a heightened sense of anticipation, but little feeling of urgency. This contact, whatever it was, was heading south at a pretty comfortable speed, and it looked like if his course and speed remained rather constant, he would come right to us. Our intent was to be there waiting for his arrival.

We shifted the internal watch-standing routine from transit where watches would be stood on a relaxed basis—typically an officer would have the watch forward for a six-hour period and then would be relieved and the next eighteen hours off. This meant that the officer stood "one in four"—one watch every four potential watch periods. Back aft we had been on "one in three"—one watch every three potential watch periods. In the forward part of the boat, the watch routine was changed to one in three, and back aft in the nuclear power plant, we changed to one in two—alternatively called "port and starboard." This meant that I would have the watch for six hours, then have six hours off as the other watch officer, Norm Emerson, stood his watch. Then I would relieve him, stand another six hours, and he would then relieve me. *That would go on for the next forty days.*

We began to copy messages on a more frequent basis as updates came in about the contact that was still heading south. These later messages classified the contact as an older Soviet-class nuclear submarine, and the locations, courses, and speeds became a bit more precise. We were well within the projected intercept area and calculated that the time of the intercept would be within the next twenty-four hours, and our sonar ears were focused sharply on sounds coming from the northeast. We adjusted our speed to about 5-6 knots and "went deep," choosing a depth above our test depth, but well below any that might be expected of our oncoming friend. We were simply lurking like a praying mantis waiting for a juicy meal. Everyone was on high alert.

The Soviet submarine headed our way was the first class of Soviet nuclear-powered attack- class submarines to be put into service. They had a streamlined torpedo-like hull, two reactors, twin propellers, and were originally designed to enter US naval operating areas and destroy ships using long range nuclear-tipped torpedoes. They were not well-equipped to serve as hunters of other submarines and had limited passive sonar capabilities, and they were considerably noisier than diesel submarines as well as the earlier American nuclear-powered submarines. We knew that we had a definite noise advantage and that it would be hard for them to detect us because of our superior quietness, but that did not mean we could take unnecessary chances. Our approach would be deliberate...and cautious. We were prepared to meet them covertly.

Our first sonar detection of our transiting friend came a couple of days later. He had changed course and was headed due south instead of toward the United States. We waited patiently as he approached, not wanting to get too close.

He was noisy – Ian Fleming's Chitty-Chitty-Bang-Bang comes to mind today, although at that time the musical was still a little over a year away from production. Our sonar gang locked in on him and we began to compute what is called a "fire control solution." That didn't mean that we were going to shoot anything at him; it meant that we were busy maneuvering to determine his exact course, speed—and, most important—range. This could easily have been done using active sonar (i.e., pinging at him and listening to the echos) but active sonar would have made him able to detect us, and that was something that we were not going to do at any cost. This was a covert operation, and we intended to keep it that way. Once we had a good solution and were confident that we had calculated his course, speed, and range accurately, we pulled in behind him, but not too close!!!

Submarines have "baffles." The baffles include the area in the water directly behind the submarine that is a blind spot where the submarine's forward hull-mounted sonar cannot hear. During normal operations, a submarine might now and then change course in either direction in order to give the forward sonar the ability to listen for any contact that might not otherwise be detected due to its being in the baffles. This was called "clearing the baffles."

We had another name for the baffle-clearing maneuver when Soviet submarines performed the same procedure—the "Crazy Ivan." When a Soviet submarine did a Crazy Ivan, he would sometimes reverse course quickly and steam down his previous track, listening for any other submarine that might be following. This was a dangerous event for us because it not only offered the chance of being detected; even more dangerous, it presented the chance of an underwater collision. For that reason, we carefully selected our depth so as not to be at the same depth of the Soviet submarine and when we detected what appeared to be a Crazy Ivan, we would immediately slow down and maneuver to provide a minimum aspect so that we would be as quiet as possible to remain undetected. We would also maneuver as necessary to ensure that we would always have a bearing rate on the submarine. (If the bearing rate was zero, that meant that we could be in danger of colliding with him.) Our submarine friend rarely did a Crazy Ivan, but did routinely clear his baffles, and we were soon able to predict almost to the minute when he would do so.

Determining a fire control solution strictly from passive sonar (i.e., listening only) is not an easy task. Assuming that the target is moving forward at a steady course and speed, the process involves maneuvering left and right for short periods of time to generate a bearing rate (the rate of change of the bearing to the target) and calculating the target's range. We followed our friend for several days on a course due south until he turned sharply and began to head out of the area where we were designated to patrol.

We followed him up as far as we could go within our operating area, and when it was certain that he was leaving the area, we slowed and let him open range until we lost contact. Then we turned around and decided to send a message indicating that we had been tracking the Soviet submarine, that he had left our area, and that we had broken off contact in accordance with our orders.

Almost immediately we received a flash message telling us to reverse course and modify our orders to permit us to expand our operating area. It didn't take the captain more than a few minutes to order us to reverse course, and head toward where we thought he was heading at flank speed. The plotting team began projecting possible positions of our friend: where he might be, given the same speed at which he had been transiting. We were on our way!

The transit at flank speed was not without excitement. As we were proceeding, sonar detected a contact dead ahead and indicated that it was a submerged submarine, possibly a US fleet ballistic submarine. We changed course slightly to avoid the contact, and passed it at high speed. As our Chief Sonarman, Owen (Coyote) Carlson remembers this encounter,

> "I happened to be on the BQS-6B Passive at the time. I don't know why the boomer was cavitating but that's what I heard first. He was about 10 degrees to starboard.

I checked his vertical angle and found it about 10 degrees up, so there was depth separation between us. Then we passed him on our starboard side; I got one more vertical angle of 18 degrees up. We were at a depth of 250 feet, so you can easily triangulate from that. He might have been at (or coming to) periscope depth that might explain the propeller cavitation that I heard, but he was CLOSE. But he was only close for less than a minute. Then we were gone."

What that submarine—or whatever it was—thought was happening, I do not know. But it must have given them a thrill. We didn't have any horn to honk as we sped by, nor could we wave. We just went ahead at full steam, hoping that we would soon catch up with our Soviet friend. When we reached the area where we thought he might be, we slowed to a more reasonable speed to increase our ability to detect our friend, and headed toward our best projected position of where he might be. After a few rather tense and anxious hours—BINGO!

Sonar reported a contact dead ahead that was very familiar. We closed the range slowly and as the signal became clearer, we had a definite classification. We had found our friend and he was heading east at the same speed. After falling in behind him at our usual range, we developed what we considered to be a solid solution and resumed our surveillance.

The Soviet submarine continued transiting to what appeared to be his operating area. For the next three weeks we tracked him through many baffle clearings and a few Crazy Ivans, and developed a plot of his movements and what we thought must be a patrol pattern. It appeared as though he was searching for an American fleet ballistic submarine, because his movements were certainly those of a search being conducted. At no time did we ever lose contact and at no time did we feel that we were in any danger or that there was even a remote chance that we had been detected.

During this period of constant tracking, we developed a couple of techniques that proved to be effective when he either cleared his baffles or changed course. These techniques (tactics) were carefully documented so that when our mission had been completed, we could share those with others that would certainly be following on similar missions in the future. The tactics were effective, safe, and would play a big part a couple of years later when I had the chance to "spread the word" to other deploying fast- attack nuclear submarines at the Naval Submarine School in Groton, Connecticut.

We didn't terminate the mission because we lost contact with our reliable friend. We deliberately broke off contact and headed home because we had fulfilled our mission. Additionally, we had been at sea for a long period of time–several weeks—and the COMSUBLANT staff had concerns about the amount of time that our food would hold up. We had learned a lot of things. We learned about where the Soviet submarine was patrolling, but the most important thing we learned was how to use this most powerful and capable submarine, its reliable nuclear power plant, and most important, the awesome detection and tracking capabilities of its sonar and weapons systems. Knowledge of this capability would be shared with the US nuclear submarine force and become part of the legacy of the Super Nuke.

I can remember being in the control room as our friend made his final turn northward to proceed on another leg of his relentless and somewhat monotonous search pattern. Captain Kelln came into the control room and assumed the conn. After carefully maneuvering to give him a minimum aspect during and after his final reversal of course from southward to northward, the captain merely stayed at a speed much slower than our friend and let him drift away. We changed course slowly toward the west, and soon he was out of range and our next step was to transit back to the United States.

It was a bit nostalgic for all of us—especially Captain Kelln—to see our reliable friend fading away to the north as we slowly headed westward. We thought about the young men on that submarine, officers and enlisted men, who were fellow submariners like us. They were doing their job for their own country and were using the same sort of training and skills that we were using. We had great respect for them; we weren't "enemies"—we never fired a shot at one another—but we were adversaries of sorts. Each of us was tasked to do a mission whose objectives were similar, but focused on the specific interests of our own countries. Although we had successfully maintained constant surveillance on them for nearly thirty days, *they had earned our complete respect.* But the time to bid farewell by disappearing to the west had come.

During the course of the extended mission, things back aft in the nuclear power plant did not change. All systems operated as designed and there were no problems, due to the high quality control of construction.

I had maintained my diet and had definitely lost weight, as had several other officers. My ability to grow a nice-looking beard, however, was a disaster. Far from having a fully formed and attractive beard, I had developed a rather spotty and haphazard growth of long whiskers here and there, to the point that after about three weeks into the mission, I simply gave up and shaved the unsightly mess off my face.

From time to time during the course of the mission, I spent time forward participating in the tracking process up front in the control room. The work being done constantly by the fire control team was to maintain a solid tracking solution of the contact, to be able to know and understand where the Soviet submarine was during one of his frequent maneuvers, and to preserve his track as he proceeded with his apparent ballistic missile submarine search.

My main interest during the times I spent forward with the fire control teams was in the plotting techniques that were being used. During submarine school we were taught several different kinds of plotting techniques that had been developed during World War II—principally a technique that required us to vary our course from time to time and generate different bearing ranges that would ultimately lead to a solution. We were fortunate in this particular mission to have a contact that was quite noisy and rather easy to track without risk of detection. It was clear, however, that this would not always be the case, especially because we knew that the Soviets were developing submarines at a breakneck pace and that these submarines, according to speculation, would be much quieter. If that were true, then it would be necessary to develop innovative ways to track the quiet ones.

I was particularly intrigued by Chief Quartermaster QMC(SS) Sharp, and how he was trying to reconstruct the track of our friend during that submarine's many maneuvers. He was using the dead reckoning tracking (DRT) plot, a piece of equipment that consisted of a glass tabletop with a mechanical mechanism underneath the glass that tracked the boat's movements over time. The cursor, a pinpoint of light, gave the position of the boat and over time would move in accordance with the boat's movements to provide a track. The position of the boat would be marked with the time every minute, along with a light pencil line from the boat along the bearing of where our friend was detected. The position of the Soviet submarine on the bearing line was marked with the time with the pencil. Over time, you could trace the movement of the *Ray* and trace the movement of the Soviet submarine relative to the *Ray*, with the distance between the two marks representing the range between the two submarines. The distance between the time marks for each submarine would be the distance the two submarines had moved during the previous minute and, given that the distance was traveled over a minute, could yield the submarine's speed. This speed was constantly checked both from sonar calculations and the indicated target speed as shown on the fire control system.

Whenever the target would "zig"—either a course change or a Crazy Ivan, the possible tracks of our friend's movements could be determined by using a "speed strip" that showed how far the submarine would move over a minute's time. The use of this technique was primarily to reconstruct the Soviet submarine's movements over time and build, for the purposes of the patrol report, a history of his track. Chief Sharp would construct this track with the use of colored pencils.

Later on, after I had conceived the idea for the SSN pre-deployment training program at the Submarine School in Groton and was conducting the program with LCDR Bruce DeMars, I refined the procedure to enable the US submarines to determine the possible track of the target submarine during a maneuver. The purpose of this technique was not only to enable the US submarine to virtually see an overhead picture of what could be happening, but to plot a "worst-case scenario" that would enable the US submarine to avoid a collision with the submarine being tracked. I called this new plotting technique the "Geographic Plot"—"GEO Plot" for short—and to this day it is a technique used (electronically these days) by our fast-attack nuclear submarines. The GEO Plot, which originated on the *Ray*, became one of the major tactical contributions of the Super Nuke to the US nuclear submarine force.

Having completed our mission, we proceeded to transit directly to Norfolk, Virginia. The transit home, at a comfortable depth and a comfortable speed, was without incident, and it gave us all a chance to catch up on some much-needed sleep as well as decompress from the tensions we constantly felt when tracking that Soviet submarine for so long.

We used the time to write the patrol report that we would submit to COMSUBLANT upon our arrival, carefully reconstructing the tracks and providing appropriate commentary about how we conducted the mission. Having shifted from the patrol status watches of port and starboard, I had considerable time to work on my submarine qualification notebook and focus on getting my dolphins (qualified in submarines).

As I reflected on the mission, I couldn't help but think about the professionalism of the entire crew in conducting the mission in accordance with our operations order. I was particularly impressed with the determination of Captain Kelln to conduct the mission safely and within the parameters of completing our objectives, but showing deliberate and cautious determination. He never seemed to be seeking personal glory, unlike some submarine skippers who both preceded and followed him. He didn't behave like a "cowboy," even though he had under his command the most modern and powerful submarine that the Navy had to offer. He operated the submarine the way it was designed to be operated, taking no chances that might compromise the mission or carelessly expose the ship and crew to unnecessary risk. He oversaw the mission completely with determined and steady leadership, while showing professional respect to fellow submariners: the Soviet submarine and her crew whose unfortunate experience would be the first encounter—unknown to them—with a Super Nuke.

We surfaced approximately sixty miles at sea and ventilated the boat. This meant that we took in the fresh at sea air to replace the artificial atmosphere in which we had been living for the past two months. I distinctly remember that rather than experiencing the cool, fresh air of the open Atlantic, it was rather muggy and, in fact, had an unpleasant odor of fish. That made me realize the high quality of the internal atmosphere provided by the oxygen generators, the CO2 scrubbers, and the CO/hydrocarbon burners. It was a nearly perfect atmosphere, and much better than the real thing. That was a surprise to me.

After surfacing, I began a new routine that endured throughout the rest of my time on the *Ray*. I began to stand the watch on the bridge of the submarine. The bridge was located on top of the forward part of the sail, where we were exposed to the ocean air and totally eliminated my tendency to seasickness. It was wonderful seeing the dolphins swimming along beside us! I wondered what they might be thinking in their own world—they must have thought that they found a huge new friend . . . they swam with us constantly and were very entertaining.

As we entered the channel into Norfolk, the realization began to sink in that we had just months from the time we had left our base until the time we were entering it. We had been gone a long time. We pulled into Pier 22 during the late morning to a small crowd of families and friends who were there to greet us. It didn't take long for the crew who had not been tasked with the watch to scramble ashore to see their families and to experience the end of a two-month draught of the firm soil. One of the first items to come aboard, as I recall, was a load of groceries—the much-needed lettuce, fresh vegetables, and real milk!

Of course, as the only nuclear-qualified bachelor on board, I had the honor of conducting the plant shutdown after the extended operation. This plant shutdown was unlike the others that I had experienced during sea trials and the shakedown cruises.

A nuclear reactor that has been operating at a significant power level for an extended period of time builds up what is called a "core history." This means that the fission products that are produced from the U235 fuel tend to build up to a point where the rate of production of these dangerous elements equals the rate of decay. These fission products are highly radioactive and, for the most part, have relatively short half-lives.

The half-life of a radioactive substance is the time it takes for the level of radioactivity of the initial volume of fission products to be reduced by half of its initial value. When these fission products decay, they give off heat that must be removed from the reactor. If the heat is not removed, it can build up to a level where it will literally melt the reactor core.

When a reactor is shut down—that is, is no longer critical—it still generates heat, which immediately after shutdown can be as high as 6-7% of the reactor's total power. This is substantial, and steps must be taken to remove this heat as the fission products with the shortest half-lives decay, which can take as long as twenty-four hours. Some of the fission products have very long half-lives, and continue to be radioactive and generate heat even for so long as the normal life of the ship. So a nuclear reactor, once having operated, will always need attention to remove the decay heat. There are many ways to safely deal with this decay heat, and eventually it becomes small enough that you do not have to worry about it.

That evening, I feasted on fresh vegetables, lettuce, and milk—then enjoyed a solid night's sleep. The next morning, I was relieved of my watch in the nuclear power plant and went ashore to our apartment. It was rather strange to be home after so long underwater, but the adjustment came quickly. That evening, Frank Spangenberg and I went to the sing-along at the Little Creek officers' club. This was an experience, too, because when I encountered some of my friends, including a couple of the North Carolina schoolteachers, I was at a loss for words, because we had been totally isolated from world news for so long. I literally had no clue about what had happened in the rest of the world for the past two months, and when I was asked what we had been doing, all I could say (and all I *did* say) was "We were operating at sea. Period." That must have been boring to hear—certainly not so exciting as a Navy flyer telling them about his landings and take-offs from an aircraft carrier—but it was fun knowing that we had done something of great value to our country in the Cold War and that perhaps someday they would know at least a little bit about our adventures.

Safely back in port with the mission definitely accomplished, it was time to get ready to go on the next mission.

The adventures of the Super Nuke had just begun. Lessons had been learned, and more were to come.

Chapter Fourteen
THE SECOND MISSION

Following the mission, we spent time at Pier 22 in Norfolk doing much-needed maintenance and minor repairs, and for the next three months conducted various training exercises as well as preparing to conduct a second mission.

My focus during this time was to get my dolphins—to get qualified in submarines. Following the Christmas holidays, Captain Kelln put on the pressure for both me and Lt. John Cox to complete our qualification by putting us on port and starboard Officer of the Deck watches. This meant that I would have the watch for one day, John the next, me the next, John the next, and this grueling rotation would last for about three weeks. The pressure worked, because by late January, both John and I became qualified in submarines. I proudly pinned on my gold dolphins!

Shortly after I became qualified in submarines, I bought a new car. My trusty green Volkswagen was a thing of the past, but a fond memory. We were in the Newport News shipyard for some minor repairs and for some reason I stopped by an Oldsmobile dealership in Newport News to check out the new "442" model.

179

As I was looking at some of the cars in the lot, they were unloading a car that looked rather unusual and quite appealing. It was silver in color, obviously like the 442 models, had a wide black stripe down each side, and a black trunk. When I asked the dealer what it was, he told me he didn't know, that it had just been shipped to them and apparently Oldsmobile hadn't made very many of them.

Upon looking closer, I discovered that it was what they called a "Hurst Olds"—a very limited model with a 455-cubic inch racing engine and a Hurst transmission. Without much thought or careful planning, I bought it on the spot for the sticker price, which was $5,500 (a lot for me in those days). The next day I picked it up from the dealer, took it out on the highway, and was blown away by its power. It reminded me somewhat of the envy I had for Don Tarquin's GTO back in nuclear power school. Now I had something like that! It was thrilling! Not long after I had purchased this beauty, I watched a TV show about muscle cars, as they were called. The Hurst Olds was referred to as a "Boulevard Bully"! I was very proud of that car and wish that I would have just packaged it away somewhere until now.

There wasn't much time to socialize, even after the port and starboard watches ended. We had time, however, to socialize with our neighbors, most of whom were naval officers and their wives. Two of the wives were on the second floor of the two-story apartment that Frank Spangenberg and I shared. They were very pleasant when we would see them coming and going. We had dinner from time to time with a fellow submariner and his wife, Laughton and Jane Smith. Laughton had graduated from

USNA a year after Frank and I, and had been recently assigned to another submarine in Division 62, the *USS Scorpion* (SSN 589). Laughton was excited because the *Scorpion* would be departing on a special operation in a few days, and it would be his first. Since his wife, Jane, was with us, we didn't speak openly about what they would be doing, but mentioned that it would be a great experience and an introduction to the silent service.

On other occasions at the Little Creek sing-along, I would encounter my USNA classmate, John Sweet, with whom I had attended many classes at the academy. Bill Harwi and his wife Gail were also a couple I knew. Bill was also a USNA graduate. Both Bill and Chuck, like Laughton, were officers aboard the *USS Scorpion*. I remember giving them a salute when they left Pier 22 on the morning of February 15, 1967—all blacked out like the *Ray* on her previous mission, silent and deadly-looking, and ready for an adventure.

We began conducting "weekly ops," meaning that every other week or so we would go to sea to fulfill some mission, such as working with anti-submarine warfare aircraft, routine training, and the like. By this time, in addition to conducting several nuclear power plant start-ups before going to sea, I had been given the Officer of the Deck watch to be on the bridge of the boat from the time we had left the pier until the time we would dive about three hours later. This was a lot of fun, not only because I could completely avoid seasickness, but because just driving the Super Nuke was a lot of fun—much like the thrill one would get in driving a Ferrari or my Hurst Olds. It was one "bad ass machine," and it was fun traveling on the surface at either full speed or near full speed with the bow of the ship plowing under the waves, and our many dolphin friends swimming alongside us. There was nothing like being up there in the fresh air, and it was even more fun because I was fortunate to have the same shipmate serving as lookout each time we went to sea: EM1(SS) Art Thompson.

Art came to the *Ray* very early in the construction period and had served many watches in the nuclear power plant, operating the Electric Plant Control Panel (EPCP).

He was a natural trainer, very smart and highly competent, fun, and totally engaging. We would spend about three hours together on the bridge going out to where we would dive, and the same amount of time coming back into Norfolk. We told many stories, had a lot of laughs, and never did run into any trouble.

Many times when we went to sea on weekly ops, we would have "riders" aboard. Generally these riders were more senior officers assigned to COMSUBLANT who were going to sea with us in order for them to keep their qualifications for submarine pay. As far as I could tell, when they rode us for a few days, they didn't contribute anything to the weekly mission; they just took up space. That seemed to me to be rather odd, because they were effectively taking a short vacation, doing virtually nothing, and the result was that they would get paid more. I thought it was sort of an abuse of the system.

Occasionally we would put to sea we would be joined by a nuisance. This nuisance was a Soviet trawler—the *Laptev*—who was definitely not fishing for cod or anything else, but was certainly interested in us. We were always alerted to the presence of the *Laptev* when we went to sea, so encountering the ship was not unexpected. We knew about incidents where the captain of the ship would draw very close to submarines and other ships that he encountered, and occasionally would even cross the bow, creating the real hazard of a collision. Most of the time, however, they appeared to be curious and interested in taking a lot of photographs.

On a few occasions, they would steam behind and cross our wake, apparently seeking to sample the water or for some unknown purpose. On one trip to sea, we were ready for that little maneuver and when it appeared that the *Laptev* was going to cross our wake, we blew number two sanitary tank. (That's the sewer tank . . . or "shit tank.") We thought that if the *Laptev* wanted to sample our wake, we would enrich his sample.

One time when we had Captain Osborne, the COMSUBLANT Chief of Staff, aboard as a rider, he was on the bridge with Art Thompson and me chatting and enjoying a cigar with us. The *Laptev* was nearby, steaming a parallel course to ours about 300 yards away. They started to signal us asking us to identify ourselves and if we needed any assistance. When this happened once before, we simply ignored him. This time, however, Captain Osborne told us to send a message by flashing light. The message was, "Fuck you!" We received no response to that message, but Captain Osborne certainly enjoyed it. He was a fun guy and the first commanding officer of a Polaris submarine.

It was winter, and while the weather in southern Virginia was not nearly so severe as that which I had experienced as a child in western North Dakota, it was cold! Adding to the twenty-degree temperature (sometimes colder) was the wind caused by our steaming along at around twenty knots. We really had to bundle up to stay warm for three hours.

Toward the end of January, we received five survival suits, apparently from COMSUBLANT, to test. These were canary-yellow suits that you could put on directly over nothing but your underwear, zip up, and go to the bridge in very cold weather. I was told that they would make you float and were intended to enable a person to survive even in the freezing water of the North Atlantic. There were two suits for each lookout, two suits for the conning officer, and one suit for the captain.

Art Thompson, Captain Kelln and I tried the suits on the way out, and they were terrific! We never felt the biting wind and thought that we had a real winner. Shortly after we had submerged and were having lunch in the wardroom, Captain Kelln told our supply officer, Jerry Keller, to get us some of those suits so we could have them on a permanent basis.

Jerry nodded and said, "Well, Captain, we will need to collect some usage data first."

The captain smiled and nodded his head.

After we had completed the weekly op and had surfaced the boat, Art and I were getting ready to go to the bridge when Captain Kelln told Jerry, who was standing in the control room, "Jerry, why don't you put on your coat and hat and head to the bridge to collect some data!"

Jerry dreaded the idea because he didn't have one of the suits, but he complied and joined Art and me on the bridge. After about an hour, the captain called him back down to the control room, apparently to ask him if he had enough data! I didn't hear what Jerry had to say, but I know that he got a terrific cold. I don't remember if the canary-yellow survival suits ever became a permanent part of our wardrobe.

In March we began getting ready for another deployment and spent time at COMSUBLANT, reading previous patrol reports as well as conducting some drills relevant to our mission during weekly ops. Captain Kelln shifted some officer jobs around and made me Operations Officer, a role that would be particularly fun during the upcoming mission because I would be in the thick of things. One of my duties would be to write the mission's patrol report.

During the course of our preparation, we had been supplied with a bundle of intelligence materials—mainly highly classified articles, photographs, information about the Soviet navy, and the like. I found them to be rather problematic to use, mainly because there were so many of them and it would have been quite difficult to have them spread out all around for easy access during the mission. So with the captain's and executive officer's approval, I got out a pair of scissors and cut out those parts of the many materials that I thought would be relevant, given the nature of our upcoming mission. I created a manual called the *"Ray* Intelligence Publication and Recognition Manual" (RIPARM for short) and carefully bundled up the documents from which it was created for later secure disposal. During the mission, the RIPARM was readily accessible and was quite useful for the conning officers who had the watch. I guess that COMSUBLANT found out about the RIPARM and liked it, because I received a nice commendation from Vice Admiral Shade for creating it.

By early April we were ready to go: loaded, locked, and ready for action. At the end of the first week in April, on April 4, 1968, we heard about the assassination of Dr. Martin Luther King, Jr. That was shocking. Over the weekend before we left, we heard about riots breaking out throughout the country. I remember being able to bring my new car to the naval base for storage. I thought that it would be safe there—that the Marines would certainly prevent any riots, should they occur in Norfolk, from spreading to the base.

I had developed a budding relationship with a very nice North Carolina schoolteacher by the name of Elizabeth. We had enjoyed many dates together—having dinner, spending some evenings at the sing along at the Little Creek officers' club, etc. Over the weekend before we were to deploy, I told her that during the deployment I would not be able to call her or communicate in any way. She didn't like that at all, and was rather adamant that I tell her where I would be going, and that it was totally inappropriate and inconsiderate of me to not tell her and to not at least check in with her once in a while. I just smiled and told her that I would be going to sea, we would be conducting operations, and that I would call her in a couple of months when we returned. She asked specifically when we would return and I told her that I didn't have a clue. She became rather angry with me for not telling her anything, and we parted on rather tense terms. When I called her when we returned about eighty days later, she told me that she had moved on to greener pastures. This was one of the hazards of being in the silent service, particularly in the world of the fast-attack nuclear submarine. You could never say anything about what you were doing, and that was totally appropriate. I'm not sure what my girlfriends thought, but that didn't matter to me. I suppose they just didn't "get it"—at least, Elizabeth didn't. I never saw her again.

Late Sunday afternoon following my rather unpleasant experience with Elizabeth, I reported on board the *Ray* to meet with some individuals who would be accompanying us on our upcoming mission. The Super Nuke looked ready to go—it was completely blacked out, lying low in the water, sleek and deadly, looking almost like an angry shark, and it made me smile. We were going to have another adventure!

My role as Operations Officer for the upcoming deployment included my previous role as Communications Officer, so I still maintained my responsibility for things going on in the "radio shack." The radio shack was on the starboard side of the boat, immediately aft of the "sonar shack" and just across the passageway from the executive officer's stateroom and the ship's office. It was a restricted area, even on the boat. The door had a sign in red that said "Restricted Area," and there was a combination lock to secure the door. I unlocked it and went inside.

Inside the radio shack were all the various radio receivers, transmitters, cryptographic equipment, typewriters, etc., necessary to conduct secure naval communications with the outside world. Additionally, we had a new piece of equipment, some of it permanently installed on the 637-class submarines and some of it additional "special" equipment brought aboard specifically for this mission.

This system, called the AN/WLR-6, was a tactical electronic warfare support system consisting mainly of specialized receivers to provide intercept, surveillance, and signal parameter analysis for threat warnings, electronic countermeasures (ECM), and other purposes. This system was not normally operated by the ship's crew, but by supplemental "riders" who brought additional specialized equipment aboard—equipment that provided specific mission-oriented capabilities and were limited in the quantities produced. The riders were often referred to as "spooks" because of the highly classified nature of their work—something talked about only in the locked radio shack, never being a part of casual conversation.

This was the first special operation where all of the AN/WLR-6 capabilities along with the mission-oriented special equipment would be used. The spooks told me that they were familiar with most of the equipment, but would have to work toward developing an integrated way to use the system once on station. My thoughts immediately turned to the development of a "tactical doctrine"—that is, how would other SSNs down the line utilize this equipment effectively? Why should they have to learn how to use it from scratch when we could easily document what we were doing and create the tactical doctrine during the mission? I put that on my "to do" list.

After meeting with the spooks, I caught a few hours' sleep and was up in time to conduct the reactor plant start-up at 0200 hours. A few hours later, we had left the pier and were heading out of port, passing the Chesapeake Bay Bridge-Tunnel connecting Virginia's Eastern Shore with the Virginia mainland at Virginia Beach east of Norfolk. Art Thompson and I had settled in on the bridge for the three-hour journey and were thankful that southeastern Virginia was experiencing a very nice spring. It was warm.

When the channel leaves the entrance to the Chesapeake Bay, it does not let you go directly east of the harbor, but rather on a southeasterly course almost parallel to the Virginia Beach coastline. From the channel a couple miles from the beach, you could easily see the hotels, resorts, etc., where tourists would be enjoying themselves while we were "operating." After passing a large red buoy to port (left) about ten miles of channel to the southeast, you cut a hard turn to port (to the left) and headed directly out to sea.

In the distance, down the channel near the large red buoy, we spotted several surface ships—perhaps half a dozen destroyers and a cruiser or two. They were making the turn at the buoy and obviously heading up the same channel that we were transiting. We would be passing each of them close aboard on our port side. After watching them with our high-powered binoculars for a few minutes, it appeared as though the surface ships, now steaming directly toward us, were getting ready to "render honors" with us.

They had obviously spotted our low but unmistakable silhouette as a US nuclear fast-attack submarine.

Rendering honors is (or was at that time) a revered custom in the US Navy. In general, the guidance for rendering honors was for ships that would pass within 600 yards of a ship displaying the flag of a senior official, or within 400 yards of a boat flying the flag or pennant of a civil official, a flag officer, or a unit commander, should render "passing honors."

The lead cruiser in the long line of surface ships seemed to be displaying the flag of an admiral, but we couldn't tell from our distance what level of admiral (two, three or four stars) was aboard. In any event, we were the junior ship and, by tradition, should render honors to the surface ship. I reached back into my memory of my days at USNA about what to do, and remembered something about sounding a bugle call, then saluting, then sounding another bugle call, and dropping the salute. The bugle might have been an optional requirement—I couldn't remember. In any event, we certainly didn't have one aboard but if we had, I could have made some sort of noises on it, given my Drum and Bugle Corps experience. Maybe "Lady of Spain"—that was one of the frequent solos I played during Navy football halftime shows.

I thought that Captain Kelln should know about the approaching surface ships, so I called down to the wardroom. "Wardroom, Bridge," I said. "Captain, a long line of surface ships — obviously the second fleet — is heading up the channel and will be passing us to port. They look as though they are getting ready for us to render honors to them. We aren't even flying a flag."

After a moment or two, Captain Kelln responded, "Aw, Charlie, just wave at 'em."

My jaw dropped. I thought that my naval career would be certainly terminated if all Art and I did was "wave at 'em." I remembered, though, that we had filed a top secret movement report—a "mod out"—and we had no hull numbers or any other means of identification. We were just a fast-moving, sleek, unmarked black sea monster plowing our way southeast with a school of dolphins swimming alongside us.

Art was ready for it. He said with a huge grin, "Wave at 'em? I can do that!" Then he began to wave with both hands, yelling, "Hi, Admiral! Hi, guys!"

I wanted to crouch down and hide, but meekly stood by as Art rendered his colorful honors to the surface ships. I raised my right hand a couple of times and offered a meek wave and salute as well. Quite a few sailors on the surface ships, however, waved back at us. *Maybe my naval career isn't over after all*, I thought.

After making the hard turn to port and steaming eastward for the next two hours, we buttoned up the bridge securely, went below, and the Super Nuke submerged.

Next stop . . . our operating area.

This was a longer transit than our first mission where we intercepted the older Soviet submarine in the north Atlantic. This time we transited northward and crossed the Arctic Circle, which enabled those of us who had not been this far north before to become certified "Blue Noses." To celebrate what was for some of us an historic event, we held a very brief informal ceremony, which consisted of issuing certified Blue Nose cards to those of us who were just entering this sacred band of brethren.

Soon we slowed and made our way quietly into our operating area with our electronic and acoustic ears wide open. Since most of the time when conducting missions of this sort we would be at periscope depth, we had visual coverage as well.

As was the practice of Captain Kelln, he was focused not only on fulfilling the objectives of our mission, but also on keeping what we were doing clandestine and avoiding detection. We patrolled cautiously and conducted various operations that were called for in our operations order. At no time did we take any unnecessary risk that might have exposed us to detection.

The RIPARM publication proved to be very valuable because of the convenience and ready access to information regarding ship recognition and the like. I spent a great deal of time with the spooks, carefully writing up the proper procedures as a tactical doctrine for the use of the AN/WLR-6 and related signal intelligence equipment. The rest of my time was spent in the control room, documenting what we were doing and writing the patrol report as things were happening.

I can understand that writing about actual things that we were doing is exciting and perhaps makes for great reading, but that is not the point of this book. During this mission, we engaged in some tense and intriguing events. Those were exciting, but the more important things that we did included serving as the first of the 637-class submarines to conduct such missions, documenting specifically the tactics we used in different circumstances, creating a more useful intelligence publication that could serve to help others conducting such missions. Additionally, we wrote the specific tactical doctrine for the amazing AN/WLR-6 electronic system, and most important, demonstrated clearly that unlike "cowboy behavior" that had been practiced by some SSN commanding officers in the past, valuable intelligence could be collected and missions accomplished without taking unnecessary risks which, if such risks caused us to be detected, would either expose the ship and crew to danger, compromise the mission, or even more important, put future SSNs in unnecessary harm's way.

On or about the 20th of May, 1968, we had completed our mission and began the slow and quiet departure from our operating area. We were beginning the long transit back to our home in Norfolk, Virginia. We had settled into a transit watch schedule when we came up to periscope depth to copy a routine message. I was in the radio shack when the message came in and was completely shocked. The message from COMSUBLANT to us told us that the *USS Scorpion* (SSN 589) was missing at sea and had apparently sunk.

I immediately took the message into Captain Kelln who, after reading it, was ashen-faced. This wasn't a report of an accident somewhere to some anonymous people; this was the loss not only of fellow submariners, but fellow submariners who were in the same submarine division, were generally tied up at the same pier at the same time, lived in the same apartment complexes, and were close personal friends as well as, in some cases, classmates. I thought of Chuck Sweet, Laughton Smith, and Bill Harwi. I thought about their wives. Did they know anything yet?

I thought about why it might have happened. Was it just an on-board accident? Was it some sort of sabotage? These are questions that are still not answered to this day to a level beyond a reasonable doubt, and it is not my purpose in this book to speculate. All that I knew was that it had simply vanished. Sunk.

Captain Kelln met with the officers in the wardroom and we discussed what to do. The captain decided to send a message giving our position and offering our assistance. Shortly thereafter we sent the message and almost immediately received a response that said in no uncertain terms, "No. Get home."

The transit back was not a happy one. I spent my time writing up the patrol report as well as the tactical doctrine for the AN/WLR-6. When we reached the point to surface, we surfaced and were met soon thereafter by an oceangoing tug that brought our division commander out to the boat to brief the captain as we transited into port.

After he came aboard, he and the captain went below and Art Thompson and I stayed on the bridge to bring the Super Nuke back into port. We weren't in a jovial mood.

When we tied up at the pier, it was an eerie feeling. There were many there to greet us— just as, I suppose, there were many there to greet the *Scorpion*...but to no avail. We had completed the patrol report and it had been signed off by Captain Kelln for distribution to COMSUBLANT. I had finished the AN/WLR-6 tactical doctrine and Captain Kelln endorsed it to send it on up the chain to the appropriate office in the Pentagon, NSA, or wherever it would wind up. After these were accomplished, the captain went ashore and I went back aft to conduct the nuclear power plant shutdown, again having to deal with a reactor that had a significant core history and significant decay heat to worry about.

The next day as I was leaving the *Ray*, I looked back and saw the boat tied up to the pier, looking exactly as she had a couple of months previously as we were getting ready to depart. She had accomplished her mission once again—had offered opportunities to develop new techniques such as the ECM tactical doctrine and fulfilling all aspects of her mission safely and without detection. Her achievements had been documented for all to see and learn about in the future. Her legacy as a Super Nuke was growing.

I had a chance to go back to our apartment where I saw Laughton Smith's wife, Jane. I didn't know what to say, but later felt a strong sense of guilt because I had arrived home safely, and Laughton had not. I suspected Jane, in her own way, would normally have felt the same.

I never felt the sense of danger at sea that we somehow might have a disaster. The Super Nuke was a marvelous machine, we operated safely and in accordance with procedure, all of the members of the crew were well qualified, and we had an extraordinarily competent commanding officer whose first priority was the safety of the crew and of the ship. After returning to Norfolk, I called my mother in South Dakota, who was relieved to know that I was safe. She told me that some of her friends had expressed concern when they heard the news about the *Scorpion* disaster, but she knew that I wasn't on the *Scorpion* and felt little or no anxiety. All she knew was that we went to sea to operate.

We all needed a rest for the next few months, and then it would be on to another adventure. The Super Nuke was scheduled to have an encounter with a new and very quiet Soviet submarine.

Chapter Fifteen
THE THIRD MISSION

The mood in Norfolk following our mission—in fact, the mood in the entire submarine force—was grim. The loss of the *Scorpion* hurt us all, not only because it reminded us of the dangers we constantly faced when submerged in a hostile environment, but because we actually knew many of the individuals who now rested at the bottom of the Atlantic.

Summer was well underway in the tidewater area, and to make necessary repairs we spent a couple of weeks in the Newport News shipyard, perched on keel blocks in a dry dock. Seeing the submarine in such a way, with all the water pumped out, really makes you appreciate the size and complexity of the machine. It looks like a huge torpedo with lots of hoses hooked up to it forward and aft. Remember that even though the submarine's reactor may be shut down, there is still heat being generated, and that heat needs to be removed. Accordingly, there has to be a source of cooling water for the ship to provide this cooling.

Other hoses and wires include those for electrical power, fresh water, and cooling for air conditioning systems. You can't believe how hot it gets inside a huge black metal object resting unprotected from the hot July/August southern Virginia sun. Maintenance and minor repair work were conducted for about two weeks in dry dock, and afterward we returned to Pier 22 at the naval base to be prepared to conduct weekly ops and do routine maintenance work and minor repairs that we could perform without the help of the shipyard.

We began to prepare for our next mission, which was tentatively scheduled for the fall of 1968. This one would be similar to our first mission: an encounter with another submarine. During our preparation for the third mission, it became clear that the next encounters with a submarine would most likely not be so easy as that with our earlier friend. The Soviets were apparently pouring a lot of money and effort into noise reduction, with the intent of making their submarines quieter. Recent encounters with such submarines proved that they were achieving much success. Those submarines were considerably more difficult not only to track, but to detect in the first place. Whatever lay in wait for us when we went back out was going to be a challenge, for sure.

Our sonar gang had learned a lot of things from our experience with our friend on our first mission and started to think about what it would be like to encounter another submarine nearly as quiet as ours. Brilliant solutions to problems often look quite simple in retrospect, and the way our sonar gang tackled the quiet submarine problem is a perfect case in point. They proved, as the pioneers of this technique, how good they really were, and their efforts and contributions to sophisticated sonar tracking and ranging techniques are one of the key parts of what made the legacy of the *USS Ray* that of a Super Nuke.

In tracking another submarine, there are three basic problems that have to be solved: 1) being able to actually detect and track the bearing of the other submarine, whether it be by listening, or by some other means; 2) being able to actually classify the other submarine — what is it? what type? etc.; 3) being able to determine the other submarine's course, speed and range. A corollary to the third problem is to determine the other submarine's course, speed, and range while he is maneuvering—such as doing a Crazy Ivan.

Our fire control technicians, led by FTC (SS) Gail Litten and FT1 (SS) Owen McCoy carefully calibrated the fire control and tracking systems to be ready for action when we deployed. Simultaneously, our sonar gang, led by STC(SS) Owen (Coyote) Carlson, ST1(SS) Bill (Doggy) Dawson, ST1(SS) Ray Stanis, and others worked tirelessly on obtaining the equipment we would need to solve these three basic problems for our upcoming patrol.

In order to understand the challenges facing us and our sonar gang, it might be helpful to give you a short description of the sonar equation. This is necessary to solve the first problem: detecting the other submarine in the first place. What if you can't actually "hear" him?

First of all, in order to conduct covert surveillance and tracking missions, a submarine uses PASSIVE sonar, i.e., it does not emit pings and listen for an echo, as you might see in a movie. Sonar that uses a ping and listens for an echo is called ACTIVE sonar.

It's how you might visualize a common bat flying around, navigating and finding mosquitoes to eat.

The problem with ACTIVE sonar is that whoever is out there can hear you, so it totally defeats the purpose of being covert; you can be easily detected. Our mission to track our noisy submarine friend during the preceding fall used PASSIVE sonar, and PASSIVE sonar would be what we would be using during our upcoming mission.

Passive sonar systems such as those on board the *Ray* listen to the sounds of the sea— those generated not only by submarines we are trying to track, but sounds produced by whales, dolphins, undersea volcanoes, shrimp, the sounds of the surface waves pounding during a storm, and the like. The ocean, underwater, can be a very noisy place.

In a sense, it's like you are in a large crowded room where there are lots of other people milling around, chatting, clinking their glasses, and whatnot. Our Soviet submarine comes into the room—to make it easy to understand, let's say that the submarine is a beautiful woman in a bright-red dress. She walks through the door on the far side of the room and strolls around the room, stopping now and then to chat with someone she knows. You want to find this woman and follow her around, knowing all the time in what direction she is walking, how fast she is walking, and exactly how far away from you she happens to be at all times. You do not want to get caught following her around.

The trouble is that you have to do this blindfolded. You can't see a thing; the bright-red dress is of no help at all. All you can do is use your ears: to listen for her sounds, pick her out from the crowd, and make those determinations about which way she is walking, how fast she is walking, and how far away she is from you completely in the dark — without peeking, and without asking anyone else. If she catches you following you around, you're in trouble.

Does that sound like a challenge? It certainly is, and in order to face that challenge successfully, you have to know a lot about the sound that is in the room. You must be ready to do a little math if you want to understand what is going on. You need to understand a "sonar equation."

The beauty of this technique is we don't even have to be able to HEAR the sound that is causing the deflection. We know it is there, and we can measure the bearing to that sound from our ship! So even if the woman dressed in a bright red dress does not emit a sound loud enough for us to actually HEAR—as Jonesy was hearing submarines in *Hunt for Red October*—we can still get a bearing to her in the crowded room.

The trouble now is as follows: While we can actually track the woman in the red dress, we don't know for sure that it is actually her. We can't see her. So how can we make sure that she is the one we are following? Generally, this is a rather easy process, IF you can hear the contact.

For the woman in the red dress, maybe she has a squeaky voice that you can detect now and then. Maybe she wears some spiked high-heeled shoes that have a distinctive "click" when she walks. Maybe she is wearing some loose bracelets that clatter now and then. There may be some distinctive sounds that help you to "classify" her as a woman in a red dress. If you have enough intelligence about women who wear red dresses, maybe you can actually tell who she is — by name. It all depends on the kinds of sounds you can hear that are different from the ambient noise in the room and make her readily identifiable. It's a tricky problem, but one that is faced by a submarine that wants to keep as quiet as possible and just listen.

A large surface ship, for example, would bang and clang and you would be able to hear the ship's propellers and, if you had the appropriate intelligence manual, actually be able to tell what ship it is almost by its hull number.

But if you're not so lucky and encounter a quiet submarine, you have to be more creative. You can't hear the banging and clanging of machinery noise, nor can you hear the submarine's propeller(s). All you can hear, if you can hear anything at all, are different "tones." The problem now is that in order to hear these tones, you have to close the range. You have to get close—and you have to really be alert. This isn't like the comfort we felt at a distant range from our first Soviet submarine friend! This might be much, much closer…and things can happen fast when you're really close!

Think about those tones for a minute. Where are they coming from? Maybe a fan that isn't so sound isolated as it should be, maybe a lube oil pump, maybe something else.

In any event, you can probably get enough sounds and tones to make a definite classification that the source of noise is definitely a submarine.

So the sonar gang has been able to detect the contact and actually classify it as a submerged submarine. Now what do they do in order to determine its course, speed, and range?

The answer is to maneuver carefully. It takes two specific maneuvers to generate a range solution—and when tracking, this is a continuous process. Each successive move is the "second move" that is used with the previous move to calculate a range. It is an iterative process, of sorts, and if the target remains on a steady course and speed for ten or fifteen minutes, a good solution can be achieved. This solution is put into the main attack director in the boat's fire control system — that is always where the best solution you have about a contact should be found. (If it is not . . . why not?)

One cannot expect when tracking another submarine that you can maintain solid contact. Occasionally you WILL lose contact. The technique to use when this occurs is to ensure that there is close coordination between sonar and the fire control system. The fire control system always has the "best solution" for the target . . . and when losing contact in sonar, we assume that the target is continuing on the same course and speed with the range the same as it was when contact is lost. The fire control system feeds a generated target track (GTT) to sonar to assist them in searching for where the target should be. This works, and we needed to do this many times during the course of the mission.

As I mentioned, when tracking, things can happen in a hurry, particularly if you're close. If the target changes course, many things happen simultaneously. The rate of change of the target's bearing will change, as will the sound level and specific tonal frequencies. That was very interesting to one of the members of our sonar gang, Bill Dawson. He provided a reason which, while simple on the surface, was brilliant–not only to explain the reason for the change in tone, but what we could learn about it!

Dawson devised a way to integrate the acoustic signals and from them during maneuvers, calculate the rate of change of range—whether the target was closing or opening. This information was invaluable in determining what a submerged contact was doing. The technique, equipment needed, and specific processes were neatly documented for future use and ultimately became a significant contribution of the *Ray* and its crew to the nuclear submarine force. The contribution was one of the major reasons behind the development of the legacy of the *Ray* as the Super Nuke.

The months of September and October 1968 were spent preparing for the upcoming mission. My responsibilities changed at that time and I assumed the duties of Main Propulsion Assistant in the Engineering Department back aft in the nuclear power plant. The purpose of this move was to enable me to focus on studying for my engineer's qualification examination, which would allow me to assume the duties as chief engineer on another submarine. Captain Kelln told me that the other submarine would most likely be a new SSN in Newport News, effectively giving me a repeat performance of my experience over the past two and a half years, but with greater responsibilities. My feelings about this were rather mixed, because I felt that I was reaching a point of burnout and that I would simply like to completely get away from the nuclear submarine world for a couple of weeks and focus on something I considered more relevant on a personal level —like just watching the sunrise and sunset someplace away from what I had been doing, however exciting that had been. A lonely cabin somewhere in the Blue Ridge mountains came to mind.

Captain Kelln was a demanding commanding officer. He had probably the greatest amount of experience in nuclear power of any naval officer, having served as chief engineer aboard a fast-attack nuclear submarine and having earned enough respect from Admiral Rickover to be picked from among all available nuclear-qualified officers—submarine and surface—to be the chief engineer with the oversight of the ship's overhaul in general, and engineering plant work in detail for the *USS Enterprise* (CV-65) at the Newport News Shipyard prior to taking command of the *USS Ray*.

His experience and knowledge made it difficult, sometimes, for certain officers to feel enough confidence to approach him about problems with equipment, etc., for which they were responsible. Captain Kelln had the ability to ask penetrating questions that got to the core of the matter and really forced an officer to ensure that he had complete knowledge of a problem and its proposed solutions before approaching the captain, simply for fear of being intimidated. This is NOT a criticism of Captain Kelln; it's just a description of how junior officers grew to anticipate and sometimes fear reporting directly to him.

A shorthand description of this evolved and can be quickly demonstrated by one particular experience of our weapons officer, Tom Rossa. Shortly before our departure on our third mission, several of us were sitting in the wardroom doing some of our administrative work when Tom walked in with a forlorn look on his face after meeting with Captain Kelln, obviously with a feeling of frustration.

"Well," Tom said with a rather frustrated look on his face, "I got three IDUs, two IJDUs, and a WHRTSA!"

> We all knew what he meant. A little translation is necessary:
>
> IDU = I don't understand.
>
> IJDU = I just don't understand.
>
> WHRTSA = Who the hell is running this ship, anyhow?

I had received many IDUs and IJDUs from Captain Kelln during the nuclear power plant construction, but I don't recall ever receiving a WHRTSA. I learned from experience that whenever you had a problem dealing with your area of responsibility, you ALWAYS did your homework before talking with Captain Kelln. That wasn't bad, and it was a lesson I learned early in the construction of the *Ray*. Captain Kelln demanded that you knew what you were talking about, had thought it through, and, in most cases, had consulted the appropriate technical manual before trying to explain what the problem was and what you were going to do about it.

Sometime between our second and third mission, I talked with Captain Kelln about our experience in the construction phase and asked him how he always knew which tough questions to ask me. They always probed deep into the problem and were excellent questions, but I was wondering just how he did it, because even though I knew that he had incredible experience and knowledge about nuclear power, I thought it was almost impossible to know the details about almost everything—particularly with a new nuclear power plant.

Captain Kelln smiled, and then told me his secret, one that I've applied again and again and have shared with many people over the past forty years since my experience on the *Ray*.

"Charlie," he said, "I always ask the third question."

I didn't know exactly what he meant, so he explained it to me.

"It's a technique I used when I was in charge of overhauling the aircraft carrier *Enterprise*. I was very intimately familiar with submarine nuclear power plants, but all I knew about the carrier power plants was what I learned about nuclear power plants in general and from what I read when preparing to take on that job."

He continued, "When someone would come to me with a problem with the plant during refueling, I let him explain what the problem was and what he recommended doing about it. I listened very carefully to what he said, and generally knew what he was talking about. Then, based on what he had said, I would ask a direct question about something relevant to the problem he had described. That was the first question."

"Then," he continued, "I would listen very carefully to his answer to my specific question. Based on what he said, I would focus on something in his answer relevant to the problem and ask another specific question about what he said. That was the second question. It got us deeper into the problem."

Then he explained, "I would again listen very carefully to the answer to my question and focus directly on something in that answer that he said. Based on that, I would ask another question that, like the others, was very specific. That was the third question."

Then he smiled and told me his secret.

"If he could answer the third question confidently in a way that I could understand it and it made sense, I could safely assume that he knew what he was talking about. If he stumbled, I would tell him that I don't understand [the IDU]. If he stumbled again, I would tell him again that I just don't understand [the IJDU]. It worked for me!"

This gave me a genuine "Aha!" moment, and I thought back to all the times when I had explained things to him, not being fully prepared. It was his way of ensuring that I had done my homework before explaining a problem to him and asking him for permission to proceed with something that I had not fully thought through. So much for IDU and IJDU. I didn't ask him what would prompt a WHRTSA!

So his advice and technique of "Ask the third question" would be something that I remembered for the rest of my life, and remained one of the best pieces of advice that I ever received—but I had to learn the hard way in order to fully appreciate it. Thinking about the advice after our conversation, I made an indelible note to myself to use that technique when and if I ever became a chief engineer—and perhaps a commanding officer —of a nuclear submarine. Captain Kelln never told me where he learned this technique, whether it originated with him or he learned it from someone else earlier in his career. I thought that he probably learned it from Admiral Rickover himself—that is something I thought the admiral would do—and, even if that wasn't the case, I gave the good admiral the credit.

The advice was pure gold.

By early November we had made all preparations for the third deployment. We had learned from recent SSN experiences that we might expect the Soviets to be deploying one or more of their newer nuclear submarines. We understood that these new submarines were much quieter than our first Soviet submarine friend nearly a year earlier, so we ensured that our sonar gang and fire control gang knew what we might face and would be prepared for virtually anything. Secretly we hoped that we would again meet up with one of the new quiet submarines, whom we discovered earlier in the year and might be ready for a deployment into the Atlantic.

On Wednesday, November 13, 1968, I conducted the normal 0200 reactor start-up for the deployment and a few hours later was on the bridge with Art Thompson taking our fully blacked-out and lethal-looking predatory submarine out to our diving area about sixty miles east of Norfolk. The trip to the diving area was uneventful except for our regular dolphin friends, who seemed to enjoy the fact that we were swimming with them again.

Of newsworthy mention, notable events that had taken place over the weekend included St. Louis Cardinal Bob Gibson winning the National League's Most Valuable Player award, the US Supreme Court striking down an Arkansas law that banned teaching evolution in public schools, John Lennon and Yoko Ono appearing nude on the cover of the *2 Virgins* album, NBA rookie Elvin Hayes scoring 54 points (a career high) against the Detroit Pistons—and, what seemed common at the time, the Soviet Union performing another nuclear test in eastern Kazakhstan.

After submerging, we headed north, anticipating again the fun transition from the Labrador Current to the Gulf Stream and the pleasant transit for the next few days to our operating area where we would lurk silently until (hopefully) we would be introduced to a new friend.

The transit was uneventful—even the reaction of the nuclear power plant crossing into the warmer ocean water of the Gulf Stream. As we approached our operating area, we began to receive messages about the detection of a submarine heading our way. Contact with this apparent transiting submarine was weak and intermittent, and we were given no classification except that it appeared to be very quiet. "Could this be a new quiet friend?" we wondered.

Over the next couple of days, we positioned ourselves on his projected track, took precautions to be extra quiet, and started listening for any trace of him. Finally, we began to receive passive sonar indications—intermittent noises and definite signal to noise deflections—that a submarine was approaching. Soon we had contact, but it was weak, and we took careful steps as we maneuvered slowly back and forth to obtain a course, speed, and range solution. Wow! Was he ever quiet! And he was going slowly, making him very tough to track.

Over the next few days we determined that our contact was indeed a very quiet submarine, but it did not appear to be transiting anywhere, just patrolling in the general area, making occasional turns but no Crazy Ivans. Our classification was that we had indeed found a new friend, although we never did have the opportunity to obtain a visual confirmation. Our new quiet friend was apparently not looking for any US submarine and exhibited no searching techniques. What his mission was for, we could not determine for certain other than perhaps it was a shakedown cruise of some kind to gain experience in operating in the open ocean rather than contrived conditions near its home port.

What we learned from this rather short three-week experience with our new friend was that these new submarines were very quiet, it was difficult to maintain contact with them, and we had to practice and try to perfect our surveillance techniques using primarily sophisticated range determination methods during maneuvers. We did not have the luxury of tracking this submarine from a comfortable distance; instead, we had to move in much closer, which posed a far greater risk not only of collision, but of possible detection. Captain Kelln wisely kept us at a safe distance and made no effort at all to resort to any "cowboy" tactics that might get us into trouble even if we risked losing contact. We were collecting information that would help future SSNs who would be definitely encountering these new and quiet machines. This was a learning experience for even the Super Nuke. While the Super Nuke was clearly superior, this new character was far more difficult to deal with.

After approximately three weeks of intermittent contact, careful close-in surveillance, and taking appropriate precautions to avoid any detection, we found our friend heading back north again, apparently having fulfilled whatever mission he was on. We sensed that he was going home, our mission had been accomplished, and we began the transit back to Norfolk. We had gained valuable intelligence, mostly about the fact that we seemed to be entering a new era: *the era of the quiet submarine.*

After a routine transit, we surfaced about sixty miles east of Norfolk, Art Thompson and I headed for the bridge, and we arrived back at Pier 22 on the 20th of December. Our patrol had lasted only about six weeks, but had been highly successful.

The period from the Christmas holidays until the end of February consisted of the last two months of my tenure aboard the Super Nuke. The time was filled with maintenance, minor repairs, and preparations for an upcoming Operational Reactor Safeguard Examination (ORSE) examination—a first-time experience for the *Ray*.

Our original pre-commissioning engineer, Lt. Reid Smith, had been transferred from the *Ray* prior to our latest deployment, and a new commanding officer, Commander John Hurt, came aboard in preparation to relieve Captain Kelln. I was focused on preparing for the ORSE examination as well as worrying about my preparations for an engineer's examination at the Bureau of Naval Reactors (NR).

Having been aboard the *Ray* for almost three full years through construction, shakedown, and three special operations had taken its toll on me. I wasn't depressed, but I found myself lacking in energy and enthusiasm, and looking forward to taking an engineer's examination and then returning to the Newport News shipyard as engineer of a new SSN did not fill me with the kind of excitement that I had experienced when leaving submarine school and going to the pre-commissioning unit of the *Ray*. In fact, I had difficulty finding the energy even to do my job and felt that I would be letting Captain Kelln and the rest of my shipmates down.

Frank Spangenberg (my apartment roommate) and I had moved out of our apartment and I was residing in the Norfolk Naval Base bachelor officers' quarters. I had a room, a bathroom, shower, and a place to crash. One morning when the alarm clock buzzed, I simply turned it off and went back to sleep—not getting up, getting into my uniform, and going to the boat. I slept until around 10:00 a.m., when I received a call from our new executive officer, LCDR J.D. Williams, asking me what I was doing and telling me to get my ass down to the ship. I met with JD and Captain Kelln, who obviously were disappointed by what I had done. It wasn't intentional, but it was definitely wrong. I told them how I felt and that I was completely "burned out." I was very disappointed with myself. Maybe I had demonstrated to everyone that "I just couldn't take it," but I knew that all I needed was a little rest and relaxation. The Law of the Navy again came to mind:

> *When a ship that is tired returneth -*
> *With signs of the sea showing plain,*
> *Men put her in dock for a season.*
> *And her speed she reneweth again.*
>
> *So shall ye if per chance ye grow weary,*
> *In the uttermost part of the sea,*
> *Pray for leave for the good of the service.*
> *As much and as oft as need be!*

This time I fully understood what that law was all about!

My next step was to have a visit with the squadron's physician, who told me it appeared that all I needed was a vacation and some time to decompress. I liked that. I thought again about the Law of the Navy, shared it with him, and he agreed with it, so I went back to the ship. I definitely did not have any sort of "nervous breakdown," nor did I have any symptoms of depression, but I was granted two weeks' leave and told that I would be transferred. I did not know it at the time, but my transfer would be to the Prospective Commanding Officers' training program at the Naval Submarine School in Groton, Connecticut.

I went on leave, going to England to spend two weeks with my sister and brother-in-law. I came back with my batteries fully charged, and began the process of transferring to the submarine school. I worried some about the fact that what had happened to me would effectively terminate my fitness for nuclear power, but that was not the case. Little did I know at that time that the fourth phase of my adventure—one that turned out to be even more exciting and eventful—was about to begin.

I was on the verge of extending the legacy of the Super Nuke! And I was fired up and ready for that challenge.

Chapter Sixteen

THE IDEA - REJECTED

It had been close to three years since I drove out of the main gate of the US Naval Base in Groton, Connecticut, leaving submarine school for adventures building and operating a fast-attack nuclear submarine. It was ironic that the *Ray* had been the last choice among 104 graduates of my submarine school class that spring, but it was first in terms of experience. I had learned a lot about a submarine, just as Commander Ken Carr had told me I would, having participated in its construction and initial operations.

I had been ordered to report to the Prospective Commanding Officer (PCO) Instructor, an ex-SSBN commanding officer and the individual charged with the responsibility of providing a six-month course of tactical instruction for individuals who had been selected for submarine command. He was assisted by two commanders, each having had command of diesel submarines and two other officers, a lieutenant commander, and a lieutenant (like me). Obviously I was the junior officer in the group. For the first week, I stayed at the home of a former shipmate on the *Ray*. Eventually I found a studio apartment about a mile north of the submarine base.

After reporting to the PCO Instructor, I was assigned to teach two courses: one being a plotting technique called the "strip plot," and the other about Soviet naval ship recognition.

The strip plot is a World War II relic that was useful for diesel submarines making a periscope approach on a ship. It is useful for ranges less than 5,000 yards (some diesel boat sailors might dispute that) and requires periodic periscope observations where the individual making the observation will mark the bearing, then call out his best guess (and it is a best guess) for the range to the target, and then give the "angle on the bow." The angle on the bow is usually a guess, which becomes more accurate as the range to the target decreases. It is the angle that is measured from the bearing to the target to the target's course.

For example, if you look at a ship through a periscope and the ship is coming directly at you, the angle on the bow is zero degrees; if the ship is going in a direction that is thirty degrees from the bearing from you to the ship and will pass on your left (port) side, then the angle on the bow is port thirty degrees. With the bearing to the target, the range to the target, and the angle on the bow, you plot the ship's position, direction, and speed on the strip plot. Doing this every minute gives you an excellent solution for your fire control system and when the target reaches the most opportune point, you can fire torpedoes and probably sink him.

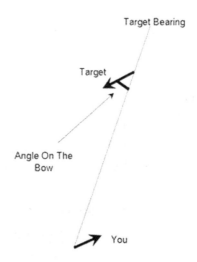

The fact is, however, that it would be very rare indeed—and, in my opinion, unwise—for a nuclear attack or Polaris submarine ever to conduct a periscope approach on any target. I began to wonder why we were teaching this particular plot to prospective commanding officers of ships that would never use it. There were other useful purposes for the equipment that was used for the strip plot.

Soviet naval ship recognition was a different matter. We had current photos of ships of every class—surface and submarines—with the exception of the most recent classes of quiet submarines. When I taught this course, I would draw sketches of the shapes of these particular submarines. As I described the Soviet surface ships and provided details about their capability, some seemed surprised that I portrayed the Soviet surface navy as being defensive in nature rather than offensive. The PCO instructor was questioning my characterization of their surface navy as well, and when I pointed out that the Soviets did not have any ships at the time, not even their most recent guided missile frigates and cruisers, that could actually project power such as the US aircraft carrier strike forces, it seemed to fall on deaf ears. Nevertheless, I continued to characterize the Soviet surface navy as defensive—because, quite frankly, *it was a defensive navy.*

Having had some time away from the *Ray* and having had time to reflect a bit on what I wanted to do with my life, I decided to resign my commission from the Navy. The principal reason was that while I was very enthusiastic about what I was doing at the time, there was no real opportunity for a social life other than hit and miss encounters here and there. Not being able to tell anyone about what I was doing was quite understandable, but it was also frustrating. It was my experience that I never had time to develop any meaningful sort of relationship, and when I did, I would only see it wither away when work or going to sea got in the way. I submitted my resignation. The PCO instructor was not too pleased with that, but I got the letter into the system and knew that it would be about a year and a half before I actually would leave the Navy. Many times during the course of the next eighteen months or so, I agonized about that decision.

I had settled in to my small one-room studio apartment and had the idea that I would work during my normal hours, and then take the time when I was free to take correspondence courses or other courses that I could find locally. I did enroll in a correspondence course from Penn State University in Fundamentals of Logic—the old "$P \rightarrow Q$" stuff. On the local level, I enrolled in the Evelyn Wood reading dynamics course, a six-week course that was supposed to increase your speed reading and comprehension. It worked for me. I was able to increase my speed reading to a level of about 2,000 - 2,500 words per minute with no loss in comprehension. The logic course worked, too, increasing my ability to understand how to put together effective arguments from facts, leading to findings, conclusions, and recommendations.

There wasn't much of a social life, except that I struck up a friendship with a divorced woman who lived in the complex, and a few young women who worked for American Airlines. We call them "flight attendants" these days, but back then they were called "stewardesses." These girls were a lot of fun and were based out of Los Angeles. They routinely took the flight from Los Angeles to Bradley Field in Hartford, Connecticut, and since Bradley was not too far from Groton, it was easy to connect with them during the many times they were in town for a layover prior to flying back. They were fun.

After about two months as a PCO tactical instructor, I had developed a good sense of the resources that the submarine school could provide to submarines that might visit the submarine base from time to time. Occasionally submarines would come into the base and sign up for a variety of training courses which, for the most part, were specialized training on various kinds of equipment for enlisted personnel. I was a bit surprised that there was no training program for officers aboard the nuclear attack submarines, given the crucial need for them to understand the kind of tactics and plotting techniques that would be required for the kinds of missions they would experience. I suppose that was a normal thing for me to wonder about, since I had just come off a submarine that had been in the thick of things—up close and personal to the best that the Soviet Union had to offer.

I decided to talk with the PCO instructor and suggest to him that we offer to teach a few tactical courses to submarine officers and crew on boats that were scheduled to deploy on special operations. I told him about the latest sonar developments and various tactics used during the course of a target maneuver to not only keep track of the target's maneuvers, but to position the submarine in a way to reduce the chances of detection or collision. I stressed the need to inform prospective commanding officers about these issues and told him that it would be quite simple to set up one of the attack centers to enable them to practice and get accustomed to this way of dealing with a more sophisticated adversary. I suggested to him that at the submarine school we had not only the resources but the recent experience to teach them.

To my surprise, his response was not only negative, but he reacted with a bit of visible disdain that I would even suggest such a thing. During the course of my conversation with him, it became clear that he had no idea what was going on in the fast-attack submarine world, he didn't seem to care or perhaps felt that any such training was unnecessary, and he obviously thought that I, as a junior officer, had nothing to offer in the way of credible ideas or meaningful suggestions. All he seemed to want to do was to steer a steady course with what the group was already doing, not "rock the boat," and put my ideas for change in some permanent storage locker.

That was a bit depressing, to say the least. But it didn't stop me.

During the next couple of weeks, I thought about this disappointing meeting and the fact that the PCO instructor had summarily dismissed my idea without so much as a "Thank you for thinking about this, I'll give it some thought," or perhaps, "That sounds like a good idea, but we just don't have the resources to do it." Those would have been acceptable answers to me, but the more I thought about what he had done, the more I thought that I would press the idea again.

The question was how to do it.

While thinking about this, I had been approached by another submarine officer who was stationed at the submarine base and worked in a different area of the submarine school. He proposed that four of us find a place in the country to rent—an old house or farm someplace—and we could share the expense. I had begun to get a little "cabin fever" living alone in a small studio apartment. It felt closed-in; it made me a bit claustrophobic, which I thought was quite odd, because for the past couple of years I had been sharing a stateroom on a submarine with two other officers, and that stateroom was about one-fifth the size of my studio apartment.

So he and I found two other officers who were interested in finding a place to live and succeeded in renting an old farmhouse, Stonecrop Farm, about ten miles east of Groton in North Stonington, Connecticut. This farmhouse was perfect for us; each of us had our own bedroom and the farmhouse itself was not only beautiful, but historic. It had been built in 1750 and featured a huge eight-foot-wide and six-foot-high fieldstone fireplace. It was ideal for four bachelors and an excellent place to hold dinner parties with our flight attendant friends from American Airlines.

Even after moving to Stonecrop Farm, the more I thought about the possibility of offering the resources of the submarine school to submarines who were preparing to deploy on special operations, and it became more intriguing. It was hopeless to talk to the PCO instructor about this, but it was also extraordinarily frustrating to be teaching tactics which were useful only if you were on a diesel submarine somewhere in the Pacific theater in World War II. I began to think about what I could do. I wanted to share my own experience with those who might benefit from it, and wanted to share what had been developed aboard the *Ray* with other submarines who would be facing the same challenges.

I was on the verge of taking a risk—one that I will never forget, and one that I would definitely repeat.

Chapter Seventeen
TAKING THE RISK

"Luck is what happens when preparation meets opportunity."

Most scholars give the credit for this quote to the Roman philosopher, Lucius Annaeus Seneca, often known as Seneca the Younger or simply Seneca. The quote resonated over and over in my mind after my idea had been rejected, and I wondered how I could turn this rejection into a stroke of luck. I had done the preparation; now all I needed was the right opportunity.

Over the course of the next few days, I was still a bit distressed about the point-blank rejection from the PCO instructor. It reminded me of an old saying that one of my shipmates on the *Ray* shared with me: *"The Navy is a hundred years' worth of tradition untouched by progress!"*

I never thought that those words were really true, because in my brief experience in the Navy, officers senior to me always took the time to listen and discuss whatever idea I had, good or bad, and showed a little respect at least for my having some creative thought.

A few days later I received the memorandum about the tactical doctrine I wrote for the electronics surveillance system—the AN/WLR-6—from Commander Bibby, who worked in the Pentagon.

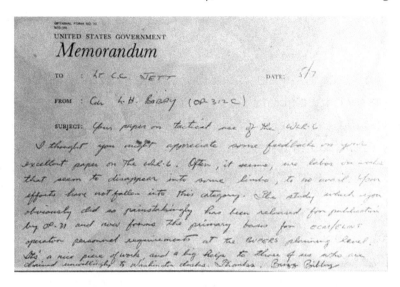

I took the memo to the PCO instructor because I thought he might like to talk about the system, or at least the subject of electronic surveillance. He was totally uninterested, looked at it for a minute, and then gave it back to me without comment. I wasn't looking for an "attaboy" or anything like that; I was just trying to initiate a discussion. Another disappointment—and it seemed to me that the old saying about progress in the Navy applied to him.

I thought about all the creative contributions that the officers and crew who were shipmates on the *Ray* had made toward advancing the submarine force's tactics, tracking techniques, and experience with more advanced Soviet submarines. I felt that it would be a disservice to the Navy if those contributions were filed away in some dust-collecting patrol report, only to be read now and then by submarine officers preparing to go on such missions. That was inefficient and a waste of all the time, effort, and money spent to develop that experience.

Most of the information was classified, and I couldn't just go out and talk to anyone about it, unless they had the appropriate security clearance. I had checked with COMSUBLANT and found that the PCO instructor had the clearance; that's why I felt comfortable talking with him. I still held the clearance as well, so I went over to the COMSUBLANT office on the submarine base and checked to see who else at the submarine school held the clearance. Fortunately, I found that the Commanding Officer of the Naval Submarine School, Captain William K. Yates, had the clearance and, most important, he had significant experience on fast-attack nuclear submarines and special operations. It would be entirely proper to have a conversation with him, because he was cleared for it... but it also meant that I would be breaking Navy protocol and going around my own boss.

Having already submitted my resignation from the Navy, however, made me think that by doing so I wouldn't really be destroying my naval career, simply because I wouldn't be in the Navy after another year and a half. So I started thinking about the possibility of privately contacting Captain Yates. On the following Saturday, I spent time writing some notes about what I might present to Captain Yates and debated over and over again whether I should contact him and what the consequences might be. From officers who served with him, I learned that he was an excellent commanding officer, an innovator, and an outstanding leader. They told me he was a good listener.

So, that Saturday afternoon, I downed a stiff drink of scotch to give me a little courage— or maybe to dampen my fears—and called Captain Yates at his home in Mystic, Connecticut, about five miles east of Groton. He answered the phone, and after introducing myself, I told him that I had something to discuss with him personally and that I didn't want to talk about it on the phone because it had to do with a matter involving special security clearances. Captain Yates suggested that I come to Mystic and visit with him at home in the afternoon the next day, a Sunday.

Early Sunday afternoon, I knocked on Captain Yates' door and he brought me into his living room. He was a very gracious man and made a concerted effort to make me feel comfortable. I told him that I felt uncomfortable speaking with him because, in effect, I was going around my boss, the PCO instructor. Captain Yates told me that our conversation would be totally private and that there would be no repercussions. He encouraged me to be candid in what I had to say, although he did not have any idea about what I was going to propose. His wife was in the room, so I politely asked him if we could go out in the back yard to discuss the matter.

In the back yard, away from any unauthorized ears, the first thing that Captain Yates wanted to know was about my experience – which boat I was on, what I had done, who was the commanding officer, and what sort of missions we had conducted.

I shared with him my enthusiasm for going to new construction of a fast-attack submarine, Captain Kenneth Carr's advice, and my experience not only in the construction period, but also the three missions. I said that I didn't want to brag, but that I had experience that was unique—the only 637-class submarine experience at the naval submarine school—and that I had experienced the new Soviet nuclear submarines up close.

I told him about the contributions to the submarine force that the *Ray* had made during its three deployments. I specifically mentioned the specific tactics that Captain Kelln had devised, the focus on conducting such missions covertly and remaining undetected, the sonar advances developed on the *Ray*, and the tactical doctrine for the AN/WLR-6. I stressed to him that these were not all MY contributions; they were the contributions of the "Super Nuke." I used that term with him, and he smiled in agreement.

I then told the captain about my idea – that the Naval Submarine School had the attack center simulators consistent with the Sturgeon-class SSN design, and that we could conduct a two-day classroom tactical training program for the officers followed by a three-day session in the simulators. Since all of the officers on a deploying SSN would have the required clearance, we would take the wraps off the tactical problems and challenge them with exactly what they would encounter during their missions. During the same time, we would make the resources of the specialized schools available for the crew.

I did not suggest that this sort of thing was something I was recommending for the prospective commanding officers, but that it was specifically for operating fast-attack submarines who were facing deployments similar to those I had experienced on the *Ray*. My argument was that it would be the best training they could possibly get before deploying, it would be less costly because the submarine would be tied up in port and not at sea, and the crew would be in port and with their families instead of being at sea for two to three weeks. I told him that the submarine school had the necessary resources to do this sort of training, that I had the appropriate tactical experience, and that we could "do the job." I suggested that with such a program we could debrief those fast-attack submarines returning from special operations and incorporate into the program the lessons that they learned and their new developments, if any—that the quality of the program would only grow as other "Super Nukes" contributed to a growing knowledge base. My suggestion was that the submarine school and the program we would create would be an up-to-date center of knowledge relevant to deploying submarines.

Captain Yates did not take long to say that he thought the concept was excellent, and thanked me for presenting it to him. He told me he agreed with me and that he felt that it was something the submarine school could and would do.

He then gave me a challenge and said, "Go down to the river [Thames] and talk to one or more of the SSN commanding officers. If one is willing to give it a try, you can do it." He told me again that our conversation was strictly confidential, and that in all matters relating to this idea, I would report directly to him. That was a relief to me, because I didn't have to go back and relate our conversation to the PCO commanding officer.

I went away from the meeting in a state of mind approaching euphoria! Those who had told me about Captain Yates were right! He listened, he thought about it, and he offered me a challenge to see if we could make it work. What was amazing to me was that he put the challenge right on my shoulders—and that was exactly what I wanted to hear. I went home and immediately put together a strategy that I would use in speaking to one or more of the commanding officers of SSNs down on the river.

The next day, I found that there were two SSNs in port, and it didn't take much thought about which one to visit. Commander Steve White was commanding officer of the *USS Pargo* (SSN650) and I had met him during the time the *Ray* was under construction in Newport News. I learned that the *Pargo* was going to deploy during the next couple of months on a special operation—and that was critical. They were a prime candidate to test the program, *if* I could get their agreement. But getting Commander White's agreement would be the *challenge of challenges*.

Steve had a reputation for being very, very smart as well as tough as nails. I knew that if I could get him to do it and he found it useful, then ANYONE would find it useful. I also knew that Steve had been the junior officer aboard the *USS Nautilus* (SSN 571) the first nuclear-powered submarine; the commanding officer of the *Nautilus* was Eugene Wilkinson, who was the current Commander Submarine Forces Atlantic (COMSUBLANT). I felt that if Steve thought it was a good idea, then Vice Admiral Wilkinson would know that fact in short order. So I went down to the *Pargo* to see Commander White.

The appearance of a Sturgeon-class SSN sitting alongside a pier has always impressed me. On the surface, the boat was about 90% submerged, and it looked like a high-tech black steel killer with the number "650" painted in white on the sail. After getting permission to come aboard, I went down below and to Captain White's stateroom. (Commanding officers of US Navy ships are always referred to as "Captain" aboard ship. So I referred to Commander White as "Captain White.")

As always, Steve was friendly but all business. "Bill Yates told me you have a proposal for me," said Steve. I swallowed hard and gave my pitch about SSN pre-deployment training as a pilot for the *Pargo* and the concept for such training in general for all deploying SSNs. Steve listened intently, and told me that he had read the patrol reports from the *Ray*'s previous three missions. He knew exactly what I was talking about and was familiar with the technical advances in sonar and the tactical developments that had been articulated in the reports. He remembered me from our encounter during the *Ray*'s construction and had a good relationship with Captain Kelln. He had certainly done his homework, and he then asked a few questions. Apparently satisfied with my answers, he then asked bluntly, "Can you deliver?"

I told him "absolutely," and that if I didn't deliver, he could "take my ass." This is a Naval Academy expression meaning that if someone is going to "take your ass" as a result of a bet or something of the sort, you had to bend over and he would whack your ass hard with an atlas. While Steve was not a Naval Academy graduate (Southern California), he understood what I meant and said, "You're on! Let's do it!"

I was in business! I left the *Pargo* and went back to my office at Sub School and called Captain Yates. When I told him about my visit with Steve White, he told me that Steve had already called him enthusiastically. "You've got a challenge on your hands now!" said Captain Yates.

I sat down and began the scheduling process: reserving the attack center simulator for three solid days, securing the classrooms, scheduling appropriate mini-schools for the *Pargo* crew, and trying to avoid the PCO commanding officer. He had apparently heard from Captain Yates that I was going to be doing a special project for him involving specific training for the *Pargo*. I told him that was correct and said little else.

Luck came my way in another fashion when Captain Yates tagged Lieutenant Commander Bruce DeMars, who had just recently become an instructor at the Naval Submarine School, to oversee the training. This was an excellent move because Bruce had considerable fast-attack submarine experience, having previously served on three fast-attack submarines, and had participated in approximately ten special operations. After the SSN pre deployment concept was explained to him, Bruce was eager to make the initial training a success. It was helpful that he had substantially more credentials than I to add the appropriate gravitas to the program. Bruce was definitely the right choice, as the next twenty years demonstrated—he rose to the rank of a four-star admiral and for eight years had Admiral Rickover's job as the Director of Naval Reactors.

The *Pargo* was on a tight schedule, so we had to schedule the training for the following week. For the first two days of wardroom officers, I covered the traditional tactics taught in the PCO school—with the exception of the antiquated strip plot—as well as a thorough review of Soviet naval ship recognition. During the afternoon of the last day, I taught some tactics that had been used during recent SSN operations involving close-in tracking of other submarines as well as techniques to approach ships on the surface for close-in observation without being detected. Although this was the first time I taught this material to anyone, Steve White commented afterward that it seemed completely in line with what he had experienced. We were then ready for three days in the attack center simulators.

During the simulations, we focused on close-in tracking of submerged submarines with minimal sonar information so they could use the tracking tactics we developed on the *Ray*. You might remember "Jonesy" in *Hunt for Red October*. In that movie and in Clancy's book, very little was divulged about current sonar capabilities and, in fact, most of what was shown in the movie was current knowledge during World War II. What the officers and plotting crew practiced during the simulations was years ahead of what was known back then. Jonesy would have been impressed.

During the third day, Captain Yates made an appearance in the attack center as Bruce and I were putting the crew of the *Pargo* through their paces. He watched the simulation for a while and chatted with Steve White, Bruce, and me. Everyone seemed very pleased with what had been done, and I had high hopes that at least Steve wouldn't "take my ass."

What followed during the next few days was amazing. Captain

Yates sent Bruce DeMars to Norfolk, Virginia to meet with Vice Admiral Wilkinson (COMSUBLANT) who, upon hearing about the proposed program, ordered all Atlantic SSNs who were to be deployed on SSN special operations to attend this training program. Bruce and I were really in business. Apparently Steve White immediately contacted his old mentor, Vice Admiral Wilkinson, and completely endorsed the program. Steve White proved to the right choice for the challenge; over the course of the next twenty years he, too, became a four-star admiral.

Apparently the old saying *"The Navy is a hundred years' worth of tradition untouched by progress!"* wasn't completely true… as I had hoped. At least there was one significant exception.

Now the challenge was different. Bruce and I had to deliver, not only for a tough commanding officer like Steve White, but for the entire Atlantic fast-attack nuclear submarine force!

The legacy of the Super Nuke was blossoming and would continue to blossom with the many contributions of other "Super Nukes" to follow!

Chapter Eighteen
SSN PRE-DEPLOYMENT TRAINING

Immediately after receiving the go ahead from COMSUBLANT, I was reassigned to the basic submarine school with an office in the main building. My days as a member of the PCO instructor team were over. The first step that Bruce and I took was to formalize the training program agenda, which included two days of tactical training for officers, followed by three days in the attack-center simulators. Then we began the training program.

Over the course of the next year, we conducted SSN Predeployment Training for every Atlantic-based SSN that was embarking on special operations missions. When an SSN would come back from a mission, we would debrief the commanding officer and watch officers, and incorporate what they learned into the attack center simulators, to prepare the next submarine for its operation.

Later, during the summer of 1969, I traveled to Pearl Harbor, Hawaii to conduct one of the programs for a fast-attack submarine in the Pacific Fleet. When I arrived at Pearl Harbor, I was met by a shipmate from the *Ray*, FTGC (SS) Owen McCoy who had participated in a lot of the development work on the

Ray. Owen (Mac) was enthusiastic about the program, and over the course of the next few months, I made a couple of presentations, conducted one training session, and attended several briefings. It seemed to me that there was a lot of enthusiasm for such training, and Mac found that out for himself—quite unlike my PCO instructor episode!

Mac had a diesel boat-trained tactical officer as officer in charge in their attack centers, but he was willing to let Mac "do his own thing." He had the leeway to conduct whatever training he wanted, and took advantage of the opportunity. For example, during one week of training for a Polaris submarine (SSBN), he provided the training that was required, with one exception: on the last day of the training, he told them, "Today you are going to become an SSN, and do some of the things that are bordering on top secret, and do some intelligence work against a modern submarine." He kept the target on a steady course and speed until the crew had a fairly good handle on the range. Then he would have the target do a baffle clear and watch the excitement rise in the crew's attitude. Near the end of the training, he explained the ranging techniques developed by ST1(SS) Dawson on the *Ray*, and had the program operator serve as the sonar operator, checking their solutions against the computers. They all caught on very nicely.

Mac didn't think too much of it until that first SSBN captain's end of training report to COMSUBPAC, stating how valuable the training was that they had received—especially the SSN addition. Over the next few months. he increased the SSN pre-deployment training to a four- or five-day cycle, twenty-four hours a day, throwing in the confusion always experienced during a watch relief, etc., and making the simulations as realistic as possible. Mac worked a twelve-hour day during those times, and shared the work load with other trainers and the tactical officer. In his own words, Mac said, "Yes, the *Ray*— the *Super Nuke*—had really got me ready to do the best training. The tactics, procedures, and techniques for conducting missions developed by the *Ray* were becoming widespread. And the beneficiaries were the submarines in the Pacific submarine force." Mac continued his naval career and eventually retired as a Master Chief Petty Officer FTCM(SS), the highest enlisted rank in the Navy.

Back in Groton, the pre-deployment program was catching on. We conducted the training for several SSNs over the course of the next few months, our knowledge base grew, and the submarine school facilities were made more available for the operating submarine force. From time to time we would visit a SSN returning from a mission, debrief them, and then take the relevant material that they discovered and applied it to future training. Since COMSUBLANT had an office on the submarine base in Groton, it was easy to have access to recent patrol reports if we did not have the opportunity to personally debrief the crews. The legacy of the *Ray* was spreading, and many other new 637-class submarines were demonstrating that they, too, were "Super Nukes."

We tried to make the simulations as realistic as possible, with periods of boredom with not much going on, and then some frenetic maneuvering by the target. We simulated very weak contact and the use of Dawson's techniques so they would gain experience recognizing and calculating solutions and coordinating the fire control systems with "sonar" which, in this case, consisted of the attack center operators.

Since we were spending so much time in the attack center simulators and polishing the tactics of a SSN in covertly tracking other submarines, I had the perfect laboratory to create new plotting techniques to assist in this process. One of the always perplexing problems was how to track a contact through a maneuver. It was a multi- faceted problem that needed close attention. The main goal was safety: to prevent a collision. The other goals were to see if we could actually understand what the contact was doing during the maneuver, and finally, we needed to reconstruct the event so that we could describe it in the patrol report if it seemed unusual. Again, though, the primary goal was SAFETY.

Over the course of the next few weeks, I modified the old "strip plot" to a new plot that I called the "Geographic Plot" (later shortened to the "Geo Plot"). It is not my intent to describe the details of how this plot works or how some adversary could possibly use it against us. Suffice it to say that the plot worked; with a little effort and training, a Geo Plotter could track another submarine in a maneuver based on passive sonar information. We could "do it" – but I won't divulge "how to do it."

Again, the reasons this plot was devised are as follows:

- The PRIMARY purpose was to use it as a safety device to enable the SSN commanding officer to avoid a collision. It was—and always should be—a safety plot;

- The secondary purpose of the plot was to enable the SSN commanding officer to obtain a bird's-eye view of what the other submarine was doing during a maneuver and thus, capture the big picture of what was happening in real time;
- The third purpose was to enable accurate reconstruction of the target's track during a maneuver.

Not everyone understood the purposes of the Geographic Plot. Not long after I called it the Geographic Plot and had conducted numerous training sessions using the technique, I was asked to ride one of the SSNs (the *Hammerhead* [SSN 663]) where we conducted some exercises with a surface ship. Apparently the thought was that the Geographic Plot was a plot to help solve the difficult passive ranging problem, *a purpose for which the plot was definitely NOT designed.* Those who understood the purpose of the plot knew that the first input to the Geographic Plot, when properly used, was the "best range to the target." That was the starting point, and the basis on which the plot was created. Knowing that this is a bit redundant, the primary purpose was the safety of the ship.

The Geographic Plot has its limitations and is subject to errors and misuse. First, the range that is put in at the beginning of a maneuver has to be a good range; everything coming from the results of the plot depends on that range being accurate. Second, the speed assumed for the target has to be accurate, because the possible target maneuvers on the Geographic Plot depend on an accurate speed. If the speed is not accurate, then the errors accumulate with each successive bearing. Third, the bearings have to be accurate. Every bearing taken, if in error, will add to the errors in the plot.

For these reasons, the best use of the Geographic Plot is as a safety plot, to keep the submarine out of danger of a collision. The ideal individual to man the Geographic Plot should be someone who has knowledge about tactics, how ships actually move and their limitations, and who is also something of an artist. The plotter should be able to discern from the increasing numbers of solutions with each successive bearing what "makes sense" and what "doesn't make sense." The most important thing that the Geographic Plotter should do is to plot a conservative "worst-case" scenario—that is, the worst situation that might endanger the ship. I suspect, although I do not know for sure, that there have been instances when overconfidence in the Geographic Plot's ability to determine a correct solution to a target's maneuver has gotten the ship into trouble. I hope that this suspicion is without merit.

Around October or so during 1969, as we were in the thick of things conducting pre- deployment training, I began to think about what to do after I left the Navy. I was having a lot of fun doing the training and working with Bruce DeMars and many of the other fine officers whom I had met, but a date in late summer of 1970 was approaching and I felt that I ought to at least have some sort of plan.

I remember having a date with someone I had casually met and during the Saturday afternoon before my date that evening, one of my USNA classmates called because he was passing through Groton. He stopped by to see me and we talked in general terms about what I was doing, and in more specific terms about what he was doing. He had gotten out of the Navy that previous summer and we talked a bit about what ex-nuclear submariners generally did after their service obligations expired. Apparently some went directly into the workforce for corporations; one of my friends went to work for IBM doing something or other . . . and my classmate from USNA, Jack, had left the Navy and was in his first year at law school.

"Most of the guys I know," said my classmate, "go to the Harvard Business School." He was a first-year student there on a two-year track to get an MBA degree. My only experience with the MBA degree had been with Jerry Keller, who had the unfortunate experience of being required to "collect data" about yellow survival suits on the bridge of the *Ray* in sub-freezing weather. (I recalled that with a smile.) I didn't know much about the Harvard Business School, but I went ahead and obtained their application, took the appropriate tests, submitted the application, and got accepted. It sort of slipped out of my mind that I would be entering an MBA program, until a few days after my release from the Navy in the summer of 1970. On reflection, that was pretty lucky career planning (if you can call it that) because the Harvard Business School is an exceptional place and the MBA credential they give graduates certainly opens doors. When I attended the school after leaving the Navy, however, I will admit that I didn't work very hard, didn't participate much, and basically treated it as a long-awaited vacation.

During the year when Bruce and I debriefed submarines coming off special operations, there were times when a new tactic was advocated, particularly when dealing with a quiet submarine as the target and when the range to the submarine was necessarily close in order to maintain contact. One such tactic advocated was as follows. Assume you are following another submarine at close range. You detect that the target is turning to port (left), so you turn to starboard yourself and speed up. The theory is that you can stay in the target's baffles and thereby not be detected.

The reality is this: Suppose the target is turning at a rate of thirty degrees per minute. At a very close range, you would have to be going at a speed far in excess of what US nuclear submarines (also Soviet) are capable of to stay in the target's baffles. Aside from that fact, you would have to accelerate quickly (probably cavitating and making a lot of noise) and get up to speed very quickly. And a turning rate of thirty degrees per minute is a relatively slow turn. Trying to plot this technique on the Geographic Plot is interesting. The tendency—if you can get up to speed—is to "spiral in," that is, close the range. Attempting to do this maneuver in the attack center simulators generally got the submarine in considerable trouble. So our conclusion was to argue against such a tactic.

By July of 1970, an officer reported to submarine school as my relief, since I was scheduled to get out of the Navy at the end of that month. During the last week, I traveled to Pearl Harbor again to conduct some training and to provide updates that we had developed, as well as to train them on the use of the Geographic Plot. Near the end of the day before the training was to end, I received a call from the submarine school telling me that I had to come back right away so that I could be processed out of the Navy. I told them to wait a couple of days so I could complete the training, but that didn't fly. So in the evening I boarded the nonstop flight from Honolulu to JFK airport in New York, and by noon the next day I was back in Groton.

I had lunch at the Officers' Club with Bruce DeMars and my old nuclear power prototype roommate, Don Tarquin. I was agonizing about the decision to get out of the Navy, and after a couple of pitchers of beer, almost came to the conclusion that I should—well, what the heck—just stay in. I went back to my apartment, took a nap, and then gave it some more thought.

I slept well that night—I don't remember if I had any dreams or not—but the next morning, bright and early, I went into the submarine school and signed all the papers necessary to leave the Navy. There were no bands, no crowds…no one. Just me. I left the building and drove out of the submarine base with only my memories.

I had a wonderful feeling of satisfaction that I had served my country honorably for the past ten and a half years and made a significant contribution to the United States Naval Nuclear Submarine Force toward winning the Cold War.

And it was a great ride!

Chapter Nineteen
REFLECTIONS

It wasn't just any Friday. It was exactly 9,529 days or twenty-six years, one month and two days since I first set my eyes on that hulking torpedo-looking shell on the shipways at the Newport News Shipbuilding and Drydock Company in Newport News, Virginia. Now she rested silently alongside the pier in Charleston, South Carolina waiting for a final farewell from many of those who worked so hard to build her. The Super Nuke was about to retire … to be deactivated. She would be decommissioned in March of 1993 and her steel recycled ten years later in Bremerton, Washington.

USS Ray (SSN-653) Inactivation Ceremony 24 July 1992.
Photo courtesy of Owen D. "Coyote" Carlson STCM(SS) USN (ret), Plank Owner *USSRay*

Many of the officers and crew of the pre-commissioning unit who worked so hard during 1966 and early 1967 to get this Super Nuke ready for adventures were in attendance as we all paid our tributes to her, gave our final salute to her and said, collectively, "Bravo Zulu." (In navy terms, "Bravo Zulu" means "Well Done!) Soon all we would have left would be our memories.

By this time, twenty-six years after first seeing the *Ray* on the shipways, many other SSNs had joined the ranks of the Super Nuke. The many 637-class submarines, along with the following 688-class submarines and the officers and crews who participated in their risky missions, all deserve the title of "Super Nuke," simply because that is what they were. They served on the front line of the Cold War . . . *and they won. Not only did they win, they kicked ass.*

Not all the missions were successful, and it has been reported in newspapers and several books that many "broke their noses" from time to time—some (including the *Ray*) from running into an uncharted sea mount and suffering considerable damage, including injuries to the crews. But we lost none of them; they never fired a shot in anger (at least during the Cold War), and some were modified to conduct missions of a special nature. They are all included in the roster of Super Nukes.

The *Ray*—the original Super Nuke—retired as one of the most highly decorated submarines in the nuclear submarine fleet. She had her share of exciting missions; other submarines may have had greater publicity about missions that were a bit more notorious or unique, but none contributed so much to the collective knowledge of the nuclear submarine force. This was due, in large part, to the fact that the *Ray* was the first operational 637-class submarine to conduct special operations. The *USS Queenfish* (SSN 651) and the *USS Sturgeon* (SSN 637) were commissioned before the *Ray*, but did not deploy on special operations for the first year or so of their service because they were used for testing and evaluation. The *Ray* was commissioned, finished shakedown cruises, and became first in line to become operational.

It is a tribute to the officers and crew of the *Ray* that they took advantage of the opportunity to be operationally first in line. They made the tactical developments, perfected the sonar innovations, created tactical doctrines for electronic and communications surveillance, created the Geographic Plotting technique, and initiated the SSN pre-deployment training programs in both the Atlantic and Pacific to share this collective knowledge with those submarines and crews who would also become "Super Nukes."

Many of the officers serving aboard nuclear fast-attack submarines rose to flag rank (admirals), including Captain Kelln, Bruce DeMars, JD Williams, and Steve White. Captain Ken Carr, who pointed me to SSN new construction, rose to the rank of vice admiral and became the Commander of the Atlantic Submarine Force (COMSUBLANT). They were outstanding officers and served their country marvelously.

Many of the enlisted crew on the *Ray* went on to distinguished careers as well. Owen (Coyote) Carlson became a master chief petty officer (STCM(SS)), and Owen McCoy also became a master chief petty officer (FTCM(SS)). They were truly professional submariners who contributed greatly to the success of the entire nuclear submarine fleet.

My respect for Admiral Rickover and the Division of Naval Reactors when I was participating in the construction of the *Ray* has only grown over the years. It was a demanding and unforgiving organization—one that was both feared and admired. Admiral Rickover was someone who was feared, but if you did your job right, you would have no problem. If you were doing your job right and you had a problem, he could be the best friend you ever had.

There were some, however, that did not fall into the admiration category. In the Navy, officers whose ambitions are to rise in the ranks and whose ambitions are particularly sensitive to their reputations are quite easy to recognize and are individuals whom the wise will not cross. Regarding these individuals, there is another Law of the Navy that applies and one that I always used:

> *Take heed what you say of your seniors,*
> *Be your words spoken softly or plain,*
> *Lest a bird of the air tell the matter,*
> *And so shall ye hear it again.*

And there is another Law of the Navy that applies and one that I always tried to follow, with the exception of my going around the PCO instructor:

> *Dost think in a moment of anger,*
> *'Tis well with thy seniors to fight,*
> *They prosper who burn in the morning,*
> *The letters they wrote overnight.*

While I was successful in getting the pre-deployment training program started and participated in its success, I made no friend of the PCO instructor and had I stayed in the Navy, think that I would have heard about it later. That was the only time I "went around" my boss—both in the Navy and later on during my career.

I would do it again!

I had the pleasure of serving with some marvelous officers—individuals with whom I've maintained contact over the years: Admiral Bruce DeMars, Admiral Steven A. White, Vice Admiral J.D. Williams, Captain William K. Yates, Rear Admiral Al Kelln, and Captain Donald C. Tarquin. They taught me good lessons—lessons that I've applied all my life, and I thank them for those lessons as well as thanking them for their service to our country.

I also had the privilege of serving with some of the finest career enlisted men in the Navy, including Chiefs Owen (Coyote) Carlson, Owen (Mac) McCoy, Gail Litten, and EM1(SS) Art Thompson. Thank you, gentlemen, for tolerating a junior officer who learned many valuable lessons from you. And thank you, Art, for the many wonderful and fun hours on the bridge of the Super Nuke.

The primary lessons I learned serving in the nuclear navy were as follows: If you are going to do something, then do it well . . . to the best of your ability. If you say you're going to do something, then do it. Whatever you do in life, do things that contribute to the betterment of your fellow man, and for your country.

The nuclear submarine force was the best organization with which I have ever had the privilege of being associated. Many individuals had problems with the strict adherence to quality and procedure in the construction, operation, and maintenance of the naval nuclear power plants. Not I. The nuclear submarine force taught me how to do things right.

I also learned from the nuclear navy and from USNA that an important thing in life to do is to be a contributor to society and to serve your country. I have great admiration for those who share that view. I have little respect or even tolerance for the "users" in our society—those who do not contribute to the betterment of our country, those who never served their country in a meaningful way, and especially those who work only to use the system to enrich themselves—to chase the almighty dollar, often at the expense of others.

When I think of those non-contributors or users, I'm reminded of something General George Patton said to his troops in his famous speech, and I'll modify it to suit my own purposes:

> *When someone is sitting by the fireside with his grandson on his knee and the youngster asks, "Grandpa, what did you do in the great Cold War against the Soviet Union?"*
>
> *Grandpa will probably have to cough and say, "Well,*

your granddaddy shoveled shit in Louisiana."

But you submariners won't have to do that.

"As someone who served in the nuclear submarine force you will be able to say, "Son, your granddaddy served on a Super Nuke...and won the Cold War."

To the *USS Ray* (SSN 653) I say proudly and with the greatest possible respect:

Thank you, old girl!"

Mission accomplished!

You served your country very well.

May you rest in peace!

Charles Cranston Jett

Plank Owner - *USS Ray* (SSN 653) - The Super Nuke Chicago, Illinois

Appendix A
ALBERT L. KELLN

Tales of a Submariner, Albert L. Kelln
The Path to Command and Success
As told to the Author, Charles Cranston Jett

My Early Submarine Experience

Following graduation from the United States Naval Academy in 1952 and commissioning, my first tour of duty was in the destroyer *USS Blue*, where I qualified as Officer of the Deck, while awaiting orders for flight school, my intention then being to get my wings and be a career pilot. However, when the expected date for my orders to Pensacola passed with no orders, followed by week after week of no orders, I finally went to my skipper and told him of my concern that my expected orders had somehow been lost. He took me aside and told me that he could not recommend me for naval aviation. His two Academy roommates had been killed while flying from carriers during World War II, and while he would gladly recommend me for submarines, he could not do so for aviation. My orders had arrived on time, but he had held them from me!

Much of our duty in the waters off Korea had been as part of the destroyer screen assigned to Task Force 77, and it was not uncommon to see pilots come back from a mission with their aircraft so badly shot up that they ditched the aircraft in the sea, usually near a destroyer, rather than try to land aboard the carrier. The frigid temperature of the sea during the winter months while we were there meant that the pilot, if he survived the ditching, had only a few minutes in which to be rescued before the cold water would claim his life.

Respecting my skipper, Commander Henry, it did not take too much of an argument from him to persuade me to change my mind about being a pilot, and I volunteered for submarines.

Upon completion of submarine school at Groton, Connecticut, I received orders to *USS Ronquil*, a Pacific Fleet boat, home-ported at San Diego. I took my bride and arrived in time to participate in *Ronquil's* work-up for deployment to the western Pacific. I had a knack for tactics and the Strip Plot, a method used by submarines at that time for determining target speed and range. Although being newly assigned to *Ronquil*, our Submarine Division Commander who was observing our training recommended that I be assigned to the battle action team that manned the conning tower at General Quarters. I felt that I was on the way to becoming a submariner!

This was wartime, and we quickly focused on our mission—surveillance of a Russian naval base. Our Navy needed to know what the Soviet Navy was doing.

A World War II veteran of submarine warfare, the skipper, Russ Medley, ordered "Rig for silent running." With course set for northern waters, *Ronquil* was ready and on its way.

Life aboard a diesel submarine was an uncomfortable and hard existence. Adding to the discomfort was the need to snorkel, and the constant cycling of the snorkel head valve atop the snorkel mast causing rapid changes in the air pressure was never a comfortable experience. The boat had a very limited capacity for making fresh water, so the officers and crew were each rationed a quart of water per week for body cleanliness and shaving. After a few weeks at sea, the odor inside the boat was not very pleasant, the only mitigating factor being our common reeking so that we somehow became accustomed to the smelly atmosphere.

The other factor in a diesel submarine was the cold. As we proceeded north, the sea water temperature continued to drop, and it was becoming very cold inside the boat. Interior heating was nonexistent, and we all wore layers of clothes, whether asleep or awake. From my bunk, I could touch the steel hull that was covered only with a thin layer of material and paint. It was not long before the inside of the hull was also covered with a thin layer of ice, noticeable when your body happened to touch the hull as you turned in your sleep. Eventually, even the water in the bilges froze into solid ice. We gained some respite by taking turns to go into the engine room when the diesel engines were running and drape our coats, clothes, and underclothes on the hot engines to dry them out and warm our bodies. The engine room odor of fuel oil became as sweet as roses to our frigid senses.

When we reached our patrol area, we found complete inactivity by the Soviet fleet. The Soviet Navy was going nowhere. But something else was happening.

At first we noticed only small cakes of ice of the 100-pound variety flowing south. We were able to avoid them in the daytime and at night we retreated to ice-free areas by going twenty to thirty miles offshore. We were unaware of a massive frontal storm in the Bering Sea that was breaking up the solid ice canopy and sending massive ice floes south into our area. When solid ice canopies break, they make a loud report, a warning to look out for floes. But we had no such warning, and were shocked when massive chunks of ice appeared in our vicinity.

The weather was usually overcast, making it impossible to see a ship at night unless its lights were on. Being so close to land, we snorkeled at night to charge the batteries instead of running on the surface. During daylight hours, we were submerged, running on the batteries. In both situations, the OOD manned the periscope. One night, without warning, the periscope was struck by a massive multi-ton slab of ice that bent the scope at a ninety-degree angle. We now had only the second periscope. After a long interval, the captain raised No.2 periscope for a look around and to evaluate our situation. Bang! No.2 scope had struck a large thick cake of ice and was also bent over. We could hear the scope filling with sea water. We were blind.

We headed south and away from hostile land, running visually undetectable on the electric motors powered by the batteries. Now the concern was whether the batteries would last until we were clear of the ice floe. We rigged for reduced power and said many silent prayers. I could sense the determination and resolve of the crew, but first we had to clear the ice floe that was testing the fabric and mettle of the crew.

We ran south at a speed that allowed the optimum balance between making distance while conserving battery life. We calculated we could run thus for about forty- eight hours before we would have to get surface air and charge the batteries, possibly revealing our position. At reduced power, everything using electricity, except the electric propulsion motors, was turned off, including the galley range. We even used flashlights rather than the normal overhead lights. In the dank cold and darkness, life was not very pleasant. We put on more clothes and waited.

Suddenly, the forward watch reported that the ice on the inside of the hull was starting to melt. A check of the sea water temperature showed it had warmed to 34 degrees Fahrenheit. We continued south another twelve hours and just at sunset, the captain stopped the boat and we went into a silent hover. With the last of our precious battery power, we started slowly pumping water ballast, and gingerly rose toward the surface. As we rose past the depth of 100 feet, we activated the passive sonar and determined that there were no ships in the vicinity. Several minutes later, we surfaced, opened the bridge hatch, and the captain and I went to the bridge. We were clear of the ice floe and no Russian radar activity was detected. The *Ronquil* had remained undetected.

Our two periscopes, bent over at ninety degrees, towered topsyturvy overhead. With no contacts detected either on sonar or on radar, we put two engines on propulsion and one on maximum battery charge, and headed for Yokosuka, Japan.

Fortunately, a warehouse at Yokosuka, filled with materiel from the war, yielded two good periscopes, and we were back in business.

I qualified in submarines shortly thereafter. Soon a call was received for junior submarine officers to volunteer for interviews with Captain Rickover for the second of its kind in nuclear propulsion training. I was accepted by the KOG (Kindy Old Gentleman), and proceeded to Groton, Connecticut for training. That experience was an 18-hour-per- day effort, as the need was great for qualified reactor operators. Soon the second class of graduates finished the rigorous course, and we proceeded to be assigned to various submarines under construction.

USS Skate to the North Pole

Fast forward to 1957. Having successfully completed nuclear power training, I was now assigned to the commissioning crew of *USS Skate*, under construction at Electric Boat in Groton, Connecticut. *Skate* was the lead boat in a series of four submarines designated as fast-attack submarines. The Cold War was heating up a bit and the Soviet Union was trying to extend its dominance over much of the world. The Soviet Navy was constructing a large nuclear-powered submarine fleet, and the numbers alone were intimidating. To counter this would require much personal sacrifice from our Navy, including long periods of time away from our families. It was our time to step up and do what had to be done. Our submarine fleet did so in ways that even today cannot be described in any detail. None of us in *Skate* knew what lay ahead. We only knew that we were trained and ready to make whatever sacrifices our duty called for.

The summer of 1958 was busy, with long days spent getting *Skate* and the crew ready for our next deployment. But we were not heading east, we were heading north. The officers were reading every book they could find that involved the history of Arctic explorations.

One piece of required information could not be found in the books, and that was the density of the openings that naturally occur at random in the Arctic ice cap. Called polynyas, the openings would be useful for surfacing in the ice cap. Getting that information would require going to the Arctic and counting any polynyas found. The group selected to go was comprised of the squadron commander, Captain Dennis Wilkinson, the executive officer of *Skate*, LCDR Nick Nicholson, LCDR Jeff Metzel, and me. A P2V patrol aircraft was selected to fly us there, since its clear plastic nose provided an excellent spot from which to photograph the ice and polynyas.

Nicholson and Metzel were the navigators. Captain Wilkinson was there to double-check the scope of preparations and submarine modifications being designed and installed by Electric Boat shipyard. And I was the ice profile data recorder and photographer.

The flight to Thule, Greenland, was uneventful, and we spent the night there. The next morning, as we flew over Baffin Bay, we could see hundreds of whales migrating south to the Atlantic Ocean, a once-in-a-lifetime sight. Flying north over the last point of land, called Alert, we saw for the first time the majestic beauty of the Arctic ice cap.

Nevertheless, it was a forbidding sight as well, since any aircraft forced to go down onto the cap would find the rough surface of the ice far from friendly for a safe landing. Our group became a bit sober when we discovered that the port aircraft engine had suddenly shut down, and the plane was slowly and decidedly losing altitude. We had been flying at 3,500 feet, so we did not have much of a cushion to work with. Much to our relief, the plane was equipped with a JATO (jet-assisted take-off) bottle under each engine, so the pilot fired up the JATOs now and then to regain altitude as needed. We turned back, and when Thule's runways appeared, we relaxed and I silently thanked God.

With the engine repaired, we once again headed north in our P2V—destination the North Pole. I took hundreds of photographs, counted the density of the polynyas, and filled my log book, and we headed back with data on our successful flight to the North Pole. *Mission accomplished!*

There were a number of special pieces of equipment being installed for the *Skate*'s journey, including newly designed underwater TV cameras, upward-beamed high-frequency ranging sonars, and mine detection sonar installed on the sub's deck to locate polynyas. Most fascinating to me was the inertial navigation system. Above 85 degrees of latitude, magnetic and gyrocompass systems are mostly useless, so the employment of inertial systems was the answer. We needed a system that could be run for days, unlike the inertial systems already in use on aircraft. Luckily, the Army had just developed such a system for their new Navajo missile and they agreed to loan us one. To make room for the nose cone from one of their missiles, containing the inertial system, we removed three sleeping bunks and installed the nose cone upside down in the crew's quarters. A hand-cranked Swedish computer was used to transform the inertial system's digital output to numerical latitude and longitude degrees.

Skate's Arctic mission was vital. In order to have a credible deterrent force, survivability of the nuclear warhead launching platforms was critical, be they aircraft, land-launched missiles, or sea-launched missiles. If we could station a few of our fleet ballistic missile submarines under the Arctic ice cap, the entire northern coast of the Soviet Union would be at risk. We were tasked to gather the data that would determine whether this new approach to protect the United States was realistic and feasible.

Skate and her crew were ready. As we sailed, I said the mariner's prayer: "O God, Thy sea is so great and my boat is so small...."

On Sunday, 9 August 1958, *Skate* reached the Arctic ice pack. The 107 members of the crew and the nine civilian experts were about to do what no man had ever done; we were going to both operate under and surface in the Arctic ice cap, a feat that would ease man's fear of this massive unknown area of the world.

Proceeding north, we found that the ice cap averaged about 14 feet in thickness, with up-ended ice called hummocks protruding down about 30 to 45 feet in spots. The ice cap is affected by both current and wind. The effects of the wind on the ice can occur at great distances from the actual location where the wind is acting, causing confusing shifting of ice—and a polynya, for example, to suddenly start closing without any warning. A submarine or ship in a polynya in that situation could be crushed by the closing ice, and indeed the wooden ships of early explorers experienced just that.

Commander Calvert, our skipper, decided to test our procedures for surfacing amid the ice cap. At the next large opening, our surfacing team went into action. We maneuvered the boat under the opening, then, allowing for current and ice movement, we carefully pumped out ballast to slowly bring *Skate* up to the surface. Other than a few polar bears whose nap we had interrupted, there was nothing but the vast expanse of ice. In just a few minutes, we had accomplished what we had spent months training to do.

The scientists gathered some data, and then we again submerged to continue north to the Pole. It was Sunday, and during the afternoon worship service, we read Psalm 139, which said in part that God was there in the uttermost parts of the sea, and His hand was leading us. We were entering an unknown world and all knew that the sea was the jealous possessor of the domain into which we had intruded, and was not forgiving if we made any mistakes.

We traveled north under the ice cap. When we were a mile from the Pole, we slowed to 5 knots, and adjusted course. At 0147 Greenwich Mean Time, on 12 August 1958, we passed directly over the North Pole. The ice canopy was solid at the Pole, and the deep ice hummocks made it unsafe to try to force our way through the ice canopy, so we circled the Pole at a few yards' distance and, in less than a few minutes, we had gone around the world! Now it was time to report our accomplishment to our superiors. While searching for a polynya, we stopped and hovered at 120 feet. Suddenly, the stillness in the control room was broken by noise from the radio room. Morse code signals were being received as clearly as if we were on the surface. In the run to the Pole, I had forgotten to test my new development of a floating wire antenna.

In the past summer, a friend from the Naval Submarine Sound Laboratory had informed me that an employee there had developed a concept in which a wire that could float on the surface might be able to receive radio signals while the submarine to which it was attached was deeply submerged. This involved encasing a flexible wire in a newly developed buoyant material. While test results had been mixed, with the help of several five-pound cans of coffee, I had persuaded the lab to fabricate one for me.

The duty radioman, while checking the antenna-switching panel, had remembered my floating wire antenna and turned on its switch to the radio speakers. I rotated the deck-mounted TV upward to the ice, and there it was; the wire was neatly coiled at the underside of the ice canopy and, without any degradation of signal, it was receiving the normal submarine broadcast from the United States. We had proved that my floating wire antenna was a way for our ballistic missile submarines, then under construction, to be in constant reception of the submarine broadcast, ensuring that the president could contact us without the necessity of our having to surface. The communication link had been the single big problem that had to be solved so that our fleet ballistic missile submarines could be ordered to release their missiles if the United States came under attack.

The skipper said, "Nice job, Kelln," and then ordered me to get on with locating a polynya. We needed to notify our superiors as to where we were, what we had done, and that we were safe. They tended to get excited and lose sleep if they thought one of our subs was in trouble. Word also had to be sent to the president; much was at stake in the Cold War, and he needed to know that the missile-carrying submarines could be relied upon in the event of an attack.

We surfaced about 40 miles from the Pole and sent our progress report. It was heard by a Navy radioman in the Philippines who took some persuading before being convinced to relay our message, we being an Atlantic Fleet unit.

For the first time ever, a submarine had surfaced in the Arctic ice pack, not once but twice, and had reached the North Pole. The *Skate*'s crew was anxious to refine our in- ice submarine surfacing procedures and gather operational data for future use by our nation. We knew we were doing something daring and historical and were determined to collect every bit of knowledge that was possible.

Cold War Special Operations

It was 1958, and the Cold War was intensifying. The Soviet Union was increasing the size of its armed forces, including the Soviet Navy. Although the possibility of a nuclear war was low, it was a possibility that could not be ignored, as Russia never let up in its effort to attain global communism. It thus became vital for our submarine force to patrol hot spots, to collect intelligence, and to be ready to strike the foe whenever our commanders so ordered. However, we had only three nuclear submarines with which to tame the bear, if called upon, so all three were heavily employed in peace-time special operations. The training for these was intense. We sharpened our tracking and torpedo approaches over and over again. Our intelligence-collecting team was likewise busy. With one last check from the squadron commander, we were off to the front lines of the Cold War. It is often said that a submarine is one of the few defense systems where we actually interact with a potential adversary during peacetime.

Here I must state that most of our missions remain highly classified even over half a century after their occurrence. Not only does this protect the information gained, it also protects those submariners who are on station today. Today's submarines not only engage in special operations, they also are capable of carrying long-range cruise and ballistic missiles that can neutralize any enemy's communications and defensive infrastructure within hours of the commencement of hostilities.

Skate and her sister submarines successfully completed their special missions such that it was evident to all, from the president on down, that the nature of naval warfare had dramatically changed with the advent of nuclear propulsion in ships and submarines. Admiral Rickover, now recognized as the father of nuclear power, had introduced a new and significant factor in the equation of world affairs.

Presidential Decision: Send *Skate* to the Arctic again, this time in the dead of winter.

Our Pentagon military planners had been elated with the *Skate*'s summer historic voyage to the Arctic. If this could be repeated in the coldest month, March, it would support continuous operation of our nuclear ballistic missile submarines on the northern shores of Russia, a strategic break-through for the USA war plans.

So *Skate* (with me once again onboard) sailed on 3 March 1959, with President Eisenhower's blessings, for the North Pole and a subsequent reconnaissance of the northern seas of Russia. The *Skate*'s crew was elated with the task and approached it, this time, with an anxious confidence of a seasoned explorer.

The day of 17 March 1959 was clear and brisk with a temperature of minus 32 degrees. At this time of the year, any hint on the sun above the horizon was rare. The *Skate* had just surfaced at the North Pole, a historic first for the United States of America and the crew of the *USS Skate* and one young Lieutenant Al Kelln. It had just been about a year before that I had flown over the North Pole and dropped a steel flagstaff with the American flag at this same spot. And now, the fact that I had just accomplished what no other Arctic explorer had ever attempted. I was the first (and probably only) person who had FLOWN OVER THE NORTH POLE, PASSED UNDER THE ICE AT THE NORTH POLE, AND NOW WAS STANDING ON THE ICE AT THE NORTH POLE.

In receiving this honor, the skipper of the *Skate*, Commander James Calvert acknowledged the awesome reality of doing this feat for the United States and he asked me, "Al, I need to know for historical reasons, do you want to be recognized henceforth as an Arctic pioneer with all its fame, or just remain as a dedicated Naval submariner?"

Without any hesitation I replied to the skipper, "Sir, I am dedicated to being the best submariner that I can be."

The skipper replied, "I expected you to say that. Carry on. And, oh yes, Good job!"

The details of the *Skate*'s saga are related in the book *Surface at the Pole* by James Calvert.

Additional Responsibility and Growth

While attending the Naval Academy, I had read the exploits of another young man of German ancestry from Fredericksburg, Texas by the name of Chester Nimitz. Admiral Nimitz would later be the senior World War II admiral who commanded the victorious naval forces in the Pacific Ocean. I always kept the memory of Admiral Nimitz as a guide and mentor, and the fact that one could accomplish the impossible if one put one's mind and energy to it. Admiral Nimitz was a humble but brilliant naval tactician. I pledged quietly to myself to become the same.

In early 1960, I was assigned to be the engineer of the submarine *Shark*, being constructed at Newport News, Virginia. I had passed all of the prerequisite training and qualifications to be the chief engineer of a nuclear-propelled submarine. I was determined to get the best-constructed submarine as I set about to learn the shipyard and its people and to train my navy engineering crew. The skipper of *Shark* was Commander John Fagan, who had been one of my instructors at Basic Nuclear Power School and was a brilliant naval officer.

The construction and commissioning of *Shark* proceeded on schedule, and all of the crew was anxious to get to sea and operate. One of the early deployments of *Shark* consisted of operating in the Mediterranean Sea and acquainting our naval forces with the capabilities of these new "nucs," as we were oft called. An unexpected and memorable operation in the Mediterranean was the receiving of the Queen and King of Greece aboard the ship and operating submerged with them at the diving controls. I had the opportunity to personally brief Queen Fredericka at her request about the *Skate*'s Arctic experiences, as she was considering construction of nuclear-powered electrical-generating plants in Greece.

Other operations of the *Shark* remain classified, except to say we received considerable front line operations and experience. I was advanced to the position of the *Shark*'s number two in command by being promoted to Executive Officer. All in all, I was learning fast and spending a lot of time at sea. The Russians were elbowing themselves into being a world naval power and it would rest on our submarine force to keep them at bay if things got out of hand. We were ready.

Soon after assignment to the executive officer position on *Shark*, I received orders to the submarine *John C. Calhoun* under construction at Newport News Shipyard. This was pleasing to me and my wife, as we would get some family time for a year or so until the ship was commissioned. I was the senior in charge of training the crew until the skippers would arrive. The ship was ready for Reactor Hot Ops when Commander Dean Axene arrived and assumed command. I was pleased with all of the training and progress that the crew had attained and was ready to settle back and just keep things perking. BUT, soon the phone rang for the skipper. The call was from Admiral Rickover late on a Friday night. His message was simple. "Kelln is detached from Calhoun and is to report as Chief Engineer of the nuclear aircraft carrier *Enterprise* on Monday morning."

When the skipper told me of these changes, I pleaded that I had no training on the *Enterprise* propulsion plant and asked Commander Axene to call Rickover and inform him of his mistake. The skipper, being new and wanting to support me, did so and called the admiral. His conservation was one of record in its brevity and went like this.

"Admiral, this is Axene and I just wanted to remind you that Kelln is not trained on any of the *Enterprise* propulsion systems." And then slowly Commander Axene hung up the phone. I rushed in to enquire about the conversation. Commander Axene stated that all the admiral said was, "If you want him qualified, then XXX XXXX XX he is qualified!" and hung up the phone. Suffice it to say, I reported to the *USS Enterprise* Monday morning for duty.

The *USS Enterprise* is a big ship, the largest nuclear-powered ship built in its time. The ship was due to head into the Mediterranean soon for a one-year deployment and finish with Operation "Sea Orbit," the world's first circumnavigation around the earth by Nuclear Task Force One. The other ships included the nuclear-propelled cruiser *Long Beach* and the nuclear-propelled frigate *Bainbridge*. I was on a fast learning curve, doing my department head job for fourteen hours of the day and learning the various systems about the ship that my department was responsible for a few hours each night. It was not long until I realized that the *Enterprise* was due for a routine shipyard repair and upgrade period as well as its first refueling of the ship's eight reactors.

I had great support from the executive officer on *Enterprise*. Upon my arrival Captain Pete Peterson, the exec, called an all-captains' conference wherein he announced that any order given by Kelln as engineer should be received as coming from himself, the captain, and be complied with. He also announced that three men—Kelln, the navigator, and the reactor officer, all having received nuclear propulsion training, would henceforth be the only in-port command duty officers. This often led to unbelievable scenes in foreign ports, where a string of captains, who were duty-officers-of-the-day for their departments, were getting briefed by me, a lieutenant commander, for the day's events and problems. I served on *Enterprise* under two commanding officers, both of whom gave me complete authority over my areas of concern. I am proud to say that I never was out of order with my seniors or took advantage of the situation.

I immediately added overhaul preps to my daily routine, and started to gather overhaul items requests from the other departments on the ship. The shipyard and navy planners agreed to meet me while in port in Italy to process the ship's overhaul requests. That done, the shipyard could start gathering materials and plans for the overhaul. That gave everyone six months to plan, which they might not otherwise have gotten. The Commander Navy Air Forces, through his material officer, gave me a secret authorization, my name only, for two million dollars for a "get well" fund if sudden overhaul problems were identified. Not even the skipper of the *Enterprise* could spend it, and this support gave me a great deal of confidence that I could get *Enterprise* back to sea in twelve months, far sooner than others had agreed upon.

In any case, the deployment was completed, and Operation "Sea Orbit" was a great success. The world's attention was centered on what and how nuclear propulsion had changed naval capabilities and warfare.

In spite of the huge shipyard and crew workloads and some major hurdles, the *Enterprise* was released from the shipyard in eleven months and a few days. Our Navy seniors were delighted. The post-overhaul sea trials went well and I was released from *Enterprise*. The ship immediately left for Viet Nam while I went off to command the *USS Ray*.

Command of the *Ray*

Now in my senior years, and at the time of this publication, I am challenged by author and friend Charlie Jett's question to me, which, in general terms is: What makes a leader? And specifically: How does a leader get the qualities that allow him to assimilate a desperate group of men into a smooth, well-oiled team with focused objectives and accomplish difficult objectives, many without reasonable solutions, into successful results?

My reply is: First, and perhaps most important, the leader must not be enamored of his personal thoughts and solutions, but open to learning. The leader is a tool of his past experiences and most of them are attained by observation of those who, in leadership positions, had successfully and steadfastly led a complex team into positive results. On the other hand, even if some of one's past leaders have been dismal failures, likewise we learn how and why. In the preceding pages, I have mentioned some of the leaders that have influenced and shaped my career as a naval officer. There are many others, most of whom have no idea of their influence and effect on my life. So, I pass these thoughts along with the prayer that others may benefit from my experiences. At least, the reader should realize that, in the majority of cases, most of our great leaders such as Rickover, Nimitz, Eisenhower, etc., were the product of dire circumstances. Therefore, we should never prejudge the capabilities of our young budding leaders, but feed them with the tools and experiences that will mature their nature into success.

As the first commanding officer of the submarine *Ray* I sincerely thank those leaders whom I have tried to emulate. The submarine *Ray* was operationally extremely successful, as acknowledged by her awards. She reflects that each crew member was a leader in his own right.

Postscript

After leaving *Ray*, Admiral Kelln was assigned a series of great Navy positions, including Head Submarine/Nuclear Power Personnel Control Branch, Trident Program Coordinator for CNO, and Commander Submarine Squadron 14, Scotland. His last Navy assignment was Deputy Director, Defense Intelligence Agency.

As a retired Admiral, Kelln was the Founder of the Naval Submarine League in 1982 and Co-Chairman of the Submarine Technology Symposium, of APL of Johns Hopkins University. He was appointed Aide-de-Camp to Governor Allen of Virginia on 4 November 1994. He has founded seven Christian Ministries which included a Christian Free Clothing Store, Healing Prayer Center, Pregnancy Resource Center, and a DVD Free Distribution Center. At 86 years of age, he is currently active in teaching about Home Solar Systems, Home Aquaponics Growing Centers and Free Summer School for school age children in various disciplines.

He and his wife, Cecily, reside in Llano, Texas.

Appendix B
NUCLEAR PHYSICS, REACTOR PRINCIPLES, AND NUCLEAR REACTOR TECHNOLOGY

Nuclear Physics, Reactor Principles, and Nuclear Reactor Technology

If you want to split atoms and use the resulting energy, you're going to have to know something about nuclear physics and how the process works. A solid knowledge of nuclear physics, reactor principles, and nuclear reactor technology is required because they are the essence of how energy can be extracted from nuclear fuel and transformed in such a way to generate electrical and mechanical power essential to the needs and survival of the boat.

A typical nuclear submarine propulsion plant is divided into two "loops": a primary loop that contains the water that passes through the nuclear reactor and through the steam generator, and a secondary loop that contains the steam generated from the heat from the primary coolant, and which goes to the engine room. The purpose of two "loops" is to separate the radioactive fluids and materials in the reactor compartment from the non-radioactive fluids and materials in the engine room.

All of the atom-splitting and creation of energy from the nuclear fuel is accomplished in the reactor compartment of the submarine. A simple diagram of a nuclear submarine power plant is shown in Figure One with the reactor compartment shown on the left. The reactor compartment contains the reactor itself, the "primary loop" containing the main coolant system, the "pressurizer" that controls the pressure of the primary loop, the "main coolant pumps" that circulate the primary coolant, and the "steam generator" where steam for the secondary loop is generated.

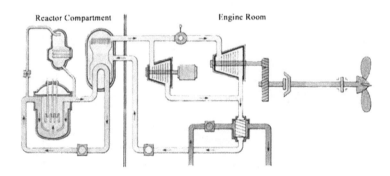

Figure One - Nuclear Submarine Propulsion Plant

Figure Two shows the reactor compartment with its important components.

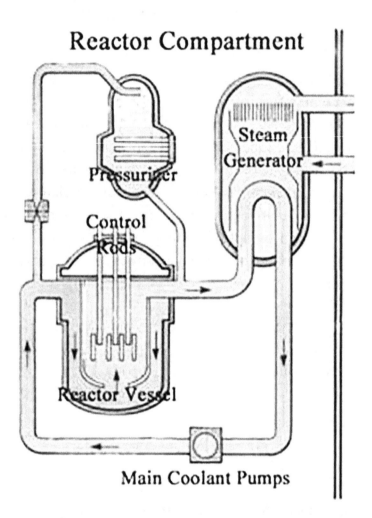

Figure Two - Reactor Compartment Detail

In order to understand the source of the energy that is extracted from nuclear fuel, you need to know a little about how the fission process works—the heart of nuclear physics.

The fuel used in naval nuclear reactors is an isotope of the heavy element Uranium 235 (U235), the same element used in atomic bombs. When a slow-moving neutron (called a "thermal" neutron) hits the nucleus of a U235 atom, the nucleus splits into two parts that fly away from one another at high speed, having a lot of kinetic energy. These parts are called "fission fragments." Radiation in the form of gamma rays is given off, as well as an average of slightly greater than two fast-moving neutrons. Gamma radiation is more powerful than X-rays and is the kind of radiation commonly used in the treatment of cancers.

What has happened when the nucleus of the atom splits apart is that some of the mass has been converted into energy. You can prove this if you measure the mass of each part that is left over after the collision and see that the sum of the remaining parts totals slightly less than mass of the original atom. Where did the missing mass go? The answer is that it was converted into energy, and you can calculate the exact amount of energy with Einstein's famous equation, $E = mc^2$.

The fast-moving fission fragments that are created when the atom splits have a lot of "kinetic energy" and this energy is converted into heat by friction created when those particles hit other atoms within the reactor's fuel cells. The result is that the fuel cells heat up and this heat can be passed along to the reactor coolant system and carried away from the core of the reactor to be used to generate steam. But since nuclear physics generally ends when the coolant leaves the reactor vessel, we will discuss the steam- generation process under the next section: Thermodynamics and Fluid Mechanics.

The energy of the gamma rays that comes when the uranium atom splits apart is not used. Lead shields have to be put around the reactor compartment to prevent individuals who are operating the reactor from becoming zapped. The energy of the fast-moving neutrons (fast neutrons) that go flying away is also not used for heat generation, but these neutrons are needed to keep the fission process going. Since U^{235} fissions almost exclusively with slow-moving neutrons (thermal neutrons), we have to find a way to slow them down so they can find other U235 atoms to split.

Slowing down neutrons can be best visualized if you imagine looking at a pool table with lots of red balls. Imagine, too, that you have a white ball that represents a neutron. If you shoot the white ball into the group of red balls, now and then the fast-moving white ball will hit one or more of the red balls. Whenever that happens, the white ball transmits some of its energy to the red balls, and the white ball slows down.

What the white ball has done is transmit some of its kinetic energy to the red balls, resulting in its slowing down. It is obvious that if you have just a few red balls on the table (a low density), then the chance of the white ball hitting the red ball is low; conversely, if you have a lot of red balls on the table (high density), the chance of a white ball hitting a red ball increases. Thus, the density of the red balls strongly affects the rate at which we can slow down the white ball. More red balls (higher density) means that the white balls will slow down more quickly.

This is what happens to the fast-moving neutrons that are generated by splitting the U235 atom. But instead of hitting red balls, the neutrons hit hydrogen atoms contained in water molecules. Water is used to slow the fast neutrons so they become thermal neutrons and are ready to split other U235 atoms. What the

NUCLEAR PHYSICS, REACTOR PRINCIPLES, AND NUCLEAR REACTOR TECHNOLOGY

water is doing is acting what is termed as a "moderator"—and this is why naval nuclear plants use what are known as "pressurized water reactors."

Keep in mind that water has "density," too. Colder water has more water molecules in a given volume, and thus more hydrogen atoms, than hotter water in the same volume. This means that if the water is colder and therefore denser, neutrons slow down and become ready to fission faster; if the water is hotter and the density is lower, the neutrons do not slow down so fast.

In order to control the population of the thermal neutrons so that the reactor can remain under control, "control rods" are used. The control rods contain an element (hafnium) that loves to absorb thermal neutrons—it just gobbles them up. Therefore, control rods are an important part of a nuclear reactor because we can control the population of the thermal neutrons and thus control the rate at which the neutrons fission the U235 atoms.

When the number of thermal neutrons being produced is exactly equal to the amount of thermal neutrons needed to create the same number of fissions, thus keeping the fission process going, we have a perfect balance in what is known as a "chain reaction" and the reactor is "critical." If the number of thermal neutrons produced exceeds the number required to keep the fission process going, the reactor is "super critical." If the number of neutrons produced is less than that needed to keep the chain reaction going, the reactor is "sub-critical."

I remember old television programs where the actors were worried that a reactor might "go critical" and somehow that was a bad thing. *It's not! In fact, it is quite normal.*

The goal for any reactor operator who wants the reactor to generate power in a steady state is to have "criticality" in the reactor core, i.e., a balanced chain reaction. Putting it another way, any nuclear reactor that is generating nuclear power in a steady state condition has a reactor that is "critical."

The energy obtained from splitting one atom is significant, but certainly not enough to power a submarine. We have to find a way to split many, many atoms in order to extract the amount of energy we actually need, but we have to be able to control the generation of this energy so that things don't get out of hand; otherwise, we would wind up with an explosion or melting down the reactor itself.

Critical reactors can operate at many levels of power, and in order to control the neutron population so that enough "fissions" are taking place so that the reactor will generate heat, we use the control rods. Control rods are used to start up the nuclear reactor and to slowly increase the power level to the point where the reactor is entering what is called the "power range"—further withdrawal of the control rods generates more heat, and the temperature of the reactor and the water moderator are increased. Control rods are, therefore, the instruments used to bring the reactor up to operating temperature so that the nuclear power plant can operate as designed.

Control rods are also used to maintain the average temperature of the nuclear reactor during operation. The average temperature of the nuclear reactor is equal to the average between the temperature of the moderating water coming into the reactor and the temperature of the moderating water that leaves the reactor with more energy. Control rods are also used as an emergency method to shut down a reactor quickly. For example,

if one or more unsafe conditions are detected, the reactor control system will automatically and rapidly insert the control rods in order to absorb the thermal neutrons and thus make the reactor go "sub-critical." When this occurs, it is known as a "reactor scram."

On a nuclear submarine, the individual who operates the reactor is called the "reactor operator" (a creative title, to be sure). It is his responsibility to operate the reactor plant control panel (RPCP) and maintain the temperature of the reactor by using the control rods.

But the reactor operator does not control the power that the reactor is generating. This is done in a more creative manner known as *"control by steam demand."* Remember that the thermal neutron population in the reactor is dependent upon two things:

- *the control rods that absorb neutrons; and*
- *the density of the moderator (water).*

This is the important principle upon which the pressurized water reactor operates.

And this is where thermodynamics and fluid mechanics enter into the nuclear propulsion plant equation.

Appendix C
THERMODYNAMICS AND FLUID MECHANICS

Thermodynamics and Fluid Mechanics

Thermodynamics is the branch of physics that is concerned with heat and temperature and their relation to energy and work. So when the nuclear reactor has finished adding energy to the coolant, thermodynamics and fluid mechanics take over to convert that energy into something useful, such as electricity, or propelling the boat through the ocean with the main turbine.

Fluid mechanics is the study of fluids (liquids, gases, plasmas) and the forces that act on them. This part of physics is important to a nuclear power plant because the energy created by the nuclear reactor moves around both the primary and secondary systems in the form of a fluid (water or steam). Further, the entire submarine and its related cooling and control systems are highly dependent on fluid mechanics: the hydraulic systems that control the boat's movements, the systems that control the boat's internal atmosphere, and the design of the submarine's shape itself, because the entire submarine exists and operates in a fluid.

Thermodynamics and fluid mechanics as taught at the nuclear power school are limited to the use and application of the energy created in the nuclear reactor—the application of both thermodynamics and especially fluid mechanics were taught at the naval submarine school, where the principles of the operating submarine (as opposed to the nuclear power plant) were in focus.

In order to understand how the energy that is created by the reactor ultimately is transferred to electrical energy and mechanical energy to propel the submarine, you need to understand the components of the secondary loop and the engine room. Figure Three shows the engine room and its important components.

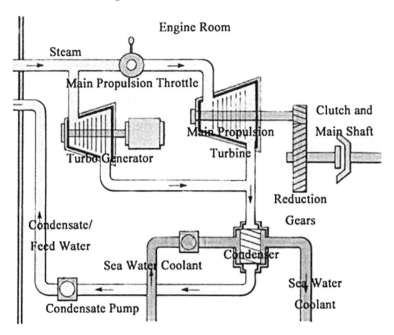

Figure Three - Engine Room Detail

Here is how it works.

Heat from the nuclear reactor is transferred by the primary coolant system to the steam generator where, obviously, steam is generated. This steam is available for use in the secondary and radioactive-free loop, to generate electricity by the turbo generators as well as to power the main propulsion turbine, which is controlled by the propulsion control panel (PCP) operator through the main throttle.

The PCP operator is in constant communication with main control room of the submarine, so when a command is given such as "All ahead one third," the PCP operator opens the main propulsion throttle and steam begins to flow to the main turbine. This steam flows through the throttle to the main turbine and expends energy to drive the main turbine. When the steam leaves the main turbine, it has lost most of that energy. Obviously we want to reuse the steam and not throw it overboard, but in order to reuse the water from the steam, the steam passes through a condenser that is cooled by sea water to convert it back to water, which can then pumped back to the steam generator. This recycled water is called "feedwater." The process of condensing the used steam back to water is the major area of inefficiency in a nuclear power plant because a lot of energy is lost—literally sent into the ocean—when the steam is condensed back into water.

Think of one gram of boiling water that is at 100 degrees Centigrade. In order to turn that one gram of water into steam, you have to add 540 calories of energy. Conversely, if you have one gram of steam at 100 degrees Centigrade, you have to take away 540 calories in order to turn that one gram of steam back into water. You can appreciate this if you think about which would be worse—being burned by one gram of boiling water at 100 degrees Centigrade, or one gram of steam at 100 degrees Centigrade. The energy in each of these choices is much different. If you are burned by one gram of steam at 100 degrees centigrade, your body will absorb 540 more calories than it would from the one gram of boiling water. In fact, after your body has been burned by the 540 calories contained in one gram of steam at 100 degrees Centigrade, you would additionally be burned by the same one gram of water now at 100 degrees centigrade. *So—be careful with steam! It has far more energy than an equivalent amount of water at the same temperature!*

This 540 calories per gram of steam is a LOT of energy, and it's called the "heat of vaporization" of water. In order to reuse the water that was used in the form of steam to run the turbines, you have to take a lot of energy away from the steam in order to convert it, and this energy is taken away by coolant water from the sea. In commercial pressurized water reactors (Westinghouse's AP1000 for example), for every 100 units of energy created by the reactor, only about 35 units can be used to generate electricity, or useful power. The other 65 units must be expelled to the environment in order to convert the steam back to reusable water. *This means that the efficiency of the AP1000 pressurized water reactor is about 35 percent.*

The feed water that is pumped back to the steam generator is colder than it was before the energy of the steam was removed, and thus it cools the steam generator. This in turn cools the primary coolant that will be pumped back into the reactor by the main coolant pumps. This primary coolant is denser than the coolant leaving the reactor (more red balls). When the colder primary coolant enters the reactor, the higher density of the moderating water slows down more neutrons, which in turn, increases the population of thermal (slow) neutrons which U235 needs to fission. This higher population of thermal neutrons generates more fission, thereby increasing reactor power to meet the steam demand. The temperature of the primary coolant leaving the reactor is increased, *but the AVERAGE temperature of the coolant leaving the reactor and that entering remains the same.*

This may sound like magic, but that is what happens. When the submarine nuclear power plant is operating, reactor power is controlled by the amount of steam needed—just what the doctor ordered! *The reactor power is controlled by steam demand, not by the reactor operator moving the control rods.*

Appendix D
RADIATION AND NUCLEAR WASTE

Nuclear physics also plays a part in what happens to a reactor when it has been operating for a period of time and builds up what is called a "core history." Remember that when fission occurs, splitting of the U235 atom generates two fission particles—these are actually atoms of different elements. When they slow down, the friction generates the heat needed to power the ship. But when they have slowed down and have come to rest… the worst may be yet to come.

These "fission fragments" can be any of a wide variety of elements and can be rather nasty critters. Some of the common fission fragment elements are cesium and iodine. Some isotopes of these elements are highly radioactive and when they decay, they give off both radiation and particles. When these particles slow down, the friction causes heat that is called "decay heat." So even when a nuclear power plant is shut down, the fission fragments that were created when the plant was operating are emitting radiation and are generating heat, *and that heat must be removed*. Failure to remove this heat can literally cause the reactor itself to melt down, such as what happened in the

Japanese nuclear power plant disaster. If these fission products escape into the atmosphere, the results can also be catastrophic, such as happened at Chernobyl.

You may be quite aware about the debate regarding nuclear power to generate commercial electricity. The debate is generally not about the ability of nuclear power to generate electricity, but about what is left over: the nuclear waste products, those nasty fission fragments that hang around for years. Until our technology is advanced enough to safely dispose of these unwanted and dangerous fission fragments, the debate will continue to rage on and on. Currently nuclear waste products are stored underground in areas such as Yucca Mountain in Nevada.

At the nuclear power school, nuclear physics is taught to enable the officer to understand not only how nuclear fuel may be used to create energy in the form of heat, but to understand the waste products of the nuclear reaction and the real dangers they present. Over the years, understanding the dangers of nuclear waste products and the dangers presented by careless operation of commercial nuclear power plants has not made me an opponent of commercial nuclear power, but certainly aware of the risks presented by such use of this powerful energy source. To me, it is much like Isaac Newton's third law of motion—for every action there is an equal and opposite reaction.

While there is not motion in the generation of nuclear power, the creation of electricity through the use of nuclear energy has both positive and negative consequences. The "action" is the creation of relatively inexpensive and usable commercial electric power; the "reaction" is the risk of nuclear accidents and, even if no accidents occur, the safe disposal and storage of spent nuclear fuel and the associated nuclear waste products.

Radioactive materials have what is known as a "half-life." In simple terms, the half-life of a radioactive element is the amount of time it takes for one-half of the quantity of that element to decay, either to a stable form, or to another radioactive element in the "decay chain." According to the Nuclear Information and Resource Service, " After ten half-lives, one thousandth of the original concentration is left; after 20 half-lives, one millionth. Generally, 10-20 half-lives is called the hazardous life of the waste. Example: plutonium-239 (Pu239, which is in irradiated fuel, has a half-life of 24,400 years. It is dangerous for a quarter million years, or 12,000 human generations. As it decays, U235 is generated; half-life: 710,000 years. Thus, the hazard of irradiated fuel will continue for millions of years. This material must be isolated from the biosphere so it will not contaminate or irradiate living things during that time." *By anyone's stretch of the imagination, this is a LONG TIME!*

If for example, a substantial amount of Pu239 contaminated the coast of Japan as a result of the Fukushima Accident in 2013, the area contaminated would continue to be dangerous for over a million years. *Unless there is technology developed to deal effectively with this disaster, the best hope of a complete clean-up would be the coming of the glaciers during the next ice age!*

Appendix E
THE SUPER NUKE PLANK OWNERS USS RAY (SSN653)

Officers

CDR Albert L. Kelln, Commanding Officer

LCDR Robert H. Harris
LCDR John R. Sopko
LT Reid H. Smith
LT Robert S. Kennedy
LT Norman P. Emerson
LT James U. Blacksher

LT John H. Cox
LT Thomas J. Rossa
LT(jg) Charles C. Jett
LT(jg) Frank A. Spangenberg, III
LT(jg) John E. Rutkowski
WO1 Charles E. Barber

Chief Petty Officers

Chief of the Boat

RMCS(SS) Robert G. Loranger
HMC Edgar D. Alvord
SKC(SS) R. E. New, Jr.
STC(SS) Steve R. Jones
ENC(SS) Merle D. Nicklas
ICC(SS) Robert E. Lambert
MMC(SS) Lee E. Bradley

TMC(SS) Lewis G. Holm

EMC(SS) William B. Burnette
STC(SS) Owen D. Carlson
RMC(SS) Edward G. Arsenault
FTC(SS) Gail A. Litten
ETC(SS) Imants Baltgalvis
QMC(SS) Warren C. Sharp
ENC(SS) George C. Lewis, Jr.

Enlisted Personnel

TM2(SS) Jackson I. Albaugh
ETR3(SS) Douglas I. Allison
MM1(SS) Gerald D. Banks
TM1(SS) Donavin D. Berg
FA(SU) Newell L. Bradley, Jr.
MM3(SS) Neil K. Carew
MM3(SU) Samual J. Clough
MM2(SS) Charles A. Dahl
EN2(SU) Richard H. Evans
ST1(SS) James E. Gabbert
ET1(SS) George F. Gillespie
TM2(SS) Arthur I. Goodwin
EM2(SS) Dennis W.W. Harris
EN1(SS) Kenneth I. Hicks
SD1(SS) Walker L. Hughes
TN(SU) Richard B. Hanson
MM2(SU) B. Kinney
MM1(SS) David M. Laird
QMSN(SS) Frank X. Lomeli
EM1(SS) Joel P. Lowman, Jr.
RM3(SU) Robert D. Martin
FTG1(SS) Owen R. McCoy
MM2(SU) Richard A. Melville
ETR2(SS) Robert B. Mhoon, III
FTG2(SU) Roy D. Miller
MM2(SU) Dennis M. Murray
MM2(SS) Wade D. Nowell
EM2(SU) James D. Olson
IC2(SS) Charles W. Ragan, II
RM1(SS) Jack Richards
ICFN(SU) James I. Sansbury
MM3(SU) Jeffrey A Schwarz
RM1(SS) Wesley R. Sisco
ETR2(SS) Thomas E. Snowdy
ST1(SS) Raymond W. Stanis
MM1(SS) Raymond G. Umstead
CS1(SS) William D. Vorlicek
ETN2(SU) Joseph R. Wallack
William V. Weissman
FN(SU) Donald W. Wilbourn

IC2(SS) Douglas R. Albright
IC2(SS) Bruce G. Arsenault
CS2(SU) Robert B. Beals
IC3(SU) Glenn S. Bland
SK1(SS) Jacob J. Bray
STS2(SS) Russell R. Chalk
TN(SU) T. W. Clarke
ST1(SS) William L. Dawson, III
EM2(SU) Edward E. Frye
MM2(SS) Arthur D. Gilbertson
YN2(SS) John E. Golz
ETN2(SS) A. L. Havens
MM2(SU) Raymond J. Hawes
STS2(SS) Curtis R. Honeycutt
ET1(SS) Jeffrey I. Hynes
FN(SU) Ellis M. Jordan
SN(SU) John C. Kohout
YN1(SS) Floyd H. Lee
ETR2(SS) Fredrick Lovret
QM1(SS) Michael R. Mandato
ETR2(SS) Joseph A McIntyr
ETR2(SS) Edward J. Mealey
MM2(SS) Samuel R. Merry
MM2(SS) Clifford S. Mickett
TM2(SS) Raymondo Morales
MM2(SS) Chris E. Norton, Jr
IC2(SS) Robert A. Oliver, Jr.
SN(SU) Johnnie R. Porter
EM1(SS) Mose T. Ramieh, Jr
SA(SU) Stanley K. Rodland
MM2(SU) Gary G. Schoeffel
ST1(SS) Donald G. Simonson
EN1(SS) Max E. Smith
IC1(SS) Charles E. Sparks
EM1(SS) Arthur S. Thompson
SN(SU) Michael A. Vituszynski
MM1(SS) Richard G Walkley
SN(SU) Alex A. Warren   ET1(SS)
QM3(SU) Frank I. Wescott
CS1(SS) Alcon Williams

ACKNOWLEDGMENTS

There are many who either inspired the writing of this book or actively participated in reviewing the chapter manuscripts, refreshing my recollections, and ensuring that the story could actually be told about fast-attack nuclear submarines without divulging any secrets or confidential information. Special thanks go to the following:

First of all, to Admiral Hyman G. Rickover, the father of the nuclear navy. He conceived and implemented the concept of the nuclear submarine and created an organization, the Division of Naval Reactors (NR), that has no peer in the world for its consistent quality and technical expertise relevant to its mission.

Next, Rear Admiral (Ret.) Albert L. Kelln, the commanding officer of the first Super Nuke, the *USS Ray (SSN653)*. I continue to address him in a way that I consider to be my highest compliment: Captain Kelln, for to me he will always be "my captain." He was the right man for the job at the time; he had the right experience, immense intelligence, and the wisdom to insist on the highest quality of performance from his crew. Captain Kelln had wisdom of caution. He was not a "cowboy," in that he knew what risks to take and always kept in mind the need to remain covert while fulfilling the requirements of a mission.

To Admiral Bruce DeMars. It is a huge understatement to say that "luck fell my way" when I started up the SSN pre-deployment training at the US Naval Submarine School in Groton, Connecticut. Bruce stepped in during the Pargo training and brought not only his broad SSN experience, but the credentials and gravitas necessary to successfully "sell" the program to the Navy brass and to actively participate in implementing the program. His efforts were a major reason that the program was not only accepted, but became highly successful in subsequent years. It took nearly fifty years, but finally what Bruce and I started is now a flag command in Groton.

To the Honorable John H. Dalton, the 70th Secretary of the Navy. John is a USNA classmate, a 50+ year close friend, and a fellow nuclear submariner. His insights and perspective as one of the nation's longest-serving Secretaries of the Navy helped to tell the story accurately while refraining from divulging classified information.

To my close friend of 50+ years and USNA classmate, Tom Elsasser, who was particularly helpful in reviewing the manuscript and providing critical and substantive suggestions. Tom lived the life of a submarine nuke for over ten years and later worked for the Nuclear Regulatory Commission for seventeen years. Thank you, Elk!

To the late Captain William Yates. Captain Yates had the leadership qualities to listen to a junior officer's proposal and the courage to "give it a try." Without him, the program would never have gotten off the ground.

To Admiral Steven A. White. Steve at the time was the commanding officer of the *USS Pargo* (SSN 650) and had the courage to be the "guinea pig" in trying the SSN pre-deployment program for the first time. He was also the toughest naval officer that I ever encountered, and that is intended to be the highest compliment I can give. He was fair, demanding, and the perfect individual to challenge the validity of a new concept.

To Vice Admiral JD Williams. JD was the executive officer on the *Ray* during our last two patrols and my "roommate." JD was an exceptional leader and motivator who had the respect of all of the officers and crew and ultimately became Commander of the United States Sixth Fleet.

To the late Vice Admiral Eugene P. Wilkinson. Admiral Wilkinson, the commanding officer of the first nuclear submarine, the *USS Nautilus* (SSN 571) was Commander of the US Atlantic Submarine Force (COMSUBLANT). He had the courage to make mandatory the SSN pre-deployment training program for all SSNs embarking on special operations.

To the crew of the *USS Ray*—the Super Nuke—my fellow "plank owners," the men who actually built the boat and participated in its commissioning. (Plank owner means they own a mythical "plank" of the ship they built.) I particularly want to put a well-deserved spotlight on four individuals. They not only participated in the construction of the Super Nuke, but also contributed significantly to my efforts in writing this book.

- STC (SS) Owen (Coyote) Carlson: Chief Carlson led a marvelous gang of sonar experts that not only demonstrated their competence during missions, but were instrumental in creating new sonar developments and techniques, which they not only conceived, but implemented.

- FTGC(SS) Gail Litten and FTG1(SS) Owen McCoy: Chief Litten and "Mac" were meticulous in setting up and perfecting the Super Nuke's fire control and tracking system, and without their competence and dedicated attention to detail, none of the missions would have been successful.

- EM1(SS) Arthur Thompson: Art was a "nuke"—an electrician—who masterfully operated the electrical plant control panel for the nuclear propulsion system. But his major achievement (to me) was being an intelligent, competent, and highly entertaining comrade on the bridge of the *USS Ray* during the many times we left our port to go to sea and return. Those times with Art are and will always be treasures for me.

To the Newport News Shipbuilding and Dry Dock Company in Newport News, Virginia. They built the very highest quality Super Nukes and were not only superb in their competence during the construction period, but also fun to work with.

Finally, to today's United States Navy. I was impressed so many years ago about the high level of quality and professionalism demanded by the submarine force and, in particular, the Division of Naval Reactors. In order to ensure that there would not be and is not any careless disclosure of still-classified material, I submitted the manuscript and experienced again the professionalism and high standards of excellence that I remembered from so many years ago. The security review took a little time, but was certainly worth the effort. Current and former submariners can rest assured that this story can be told without compromising the legacy of the "silent service" or compromising national security.

Charles Cranston Jett
Chicago, Illinois

CPSIA information can be obtained
at www.ICGtesting.com
Printed in the USA
BVHW041530010820
585164BV00008B/227